PROCESS PIPING DESIGN HANDBOOK

Volume Two
Advanced Piping Design

PROCESS PIPING DESIGN HANDBOOK

Volume One: The Fundamentals of Piping Design

Volume Two: Advanced Piping Design

PROCESS PIPING DESIGN HANDBOOK

Volume Two
Advanced Piping Design

Rutger Botermans and Peter Smith

Gulf Publishing Company
Houston, Texas

Process Piping Design Handbook
Volume Two: Advanced Piping Design

Gulf Publishing Company
2 Greenway Plaza, Suite 1020
Houston, TX 77046

10 9 8 7 6 5 4 3 2 1

ISBN-10: 1-933762-18-7
ISBN-13: 978-1-933762-18-0

Library of Congress Cataloging-in-Publication Data

Smith, Peter.
  Process piping design handbook.
      v. cm.
  Includes bibliographical references and index.
  Contents: 1. The fundamentals of piping design / Peter
Smith -- 2. Advanced piping design / Rutger Botermans and
Peter Smith.
  ISBN-13: 978-1-933762-04-3 (v. 1 : acid-free paper)
  ISBN-10: 1-933762-04-7 (v. 1 : acid-free paper)
  ISBN-13: 978-1-933762-18-0 (v. 2 : acid-free paper)
  ISBN-10: 1-933762-18-7 (v. 2 : acid-free paper)
 1. Pipelines—Design and construction. 2. Piping—Design
and construction. 3. Piping—Computer-aided design. 4.
Petroleum refineries—Equipment and supplies. I. Botermans,
Rutger. II. Title.
  TA660.P55S65 2007
  621.8'672—dc22
                                        2006038256

Printed in the United States of America
Printed on acid-free paper. ∞
Text design and composition by TIPS Technical Publishing, Inc.

This book is dedicated to my son Stewart,
who is a constant source of inspiration and motivation to me.
He is sadly missed by his family and friends,
but he is never far away.

Stewart Smith, Musician
March 11, 1972, to October 21, 2005

Contents

List of Figures

only if reverse burning is required. While pass 1 is being decoked, steam is injected into pass 2 to keep the tubes cool.   132

List of Tables

Foreword

In Volume One of the Process Piping Design Handbook, Peter Smith delivers a comprehensive and in depth instruction in the practice of piping design. He covers many important topics in exhaustive detail, from codes and standards to piping components to design practices and processes to mechanical equipment to piping materials. Volume One addresses the fundamentals of each topic exhaustive detail, creating a very strong foundation to build upon. Armed with the fundamentals, we are ready to move on to the next level. And so it is with Volume Two: Advanced Piping Design—Rutger Botermans, with the help of Peter Smith, takes us to that next level by integrating the many detailed fundamentals covered in Volume One into a broader system view of plant layout and piping design.

Rutger and Peter focus on the practice of piping design, not the tools. The volume begins with the best practices for overall plant layout, covering location of critical process equipment and its interconnecting piping systems. The best practices incorporate the many considerations required to arrive at a suitable plant layout. These considerations include process function, safety, accessibility, construction, maintenance, economics, and even aesthetics. In the subsequent chapters, Rutger and Peter walk us through every type of plant equipment—pumps, heat exchangers, tanks, and columns—explaining the various types and functions of each equipment category, the applicable codes and standards, guidelines for the design of the connecting piping, and many other design considerations. They also provide similar best practices for the design of cooling towers, pipeways, and relief systems.

For most piping design professionals, the fundamentals and best practices for piping design were learned through years of experience.

However, as anyone currently engaged in the process plant industry can attest, experienced piping design professionals are becoming harder and harder to find. Rutger Botermans and Peter Smith perform a valuable service to the piping design profession through their comprehensive, professional, and pragmatic guide to piping design.

—A. B. Cleveland Jr.
Senior Vice President
Bentley Systems, Incorporated

Preface

In Volume 1, The Fundamentals of Piping Design, the objective was to arm the reader with the basic "rules" for the design, fabrication, installation, and testing of process and utility piping systems for oil and gas refineries, chemical complexes, and production facilities at both offshore and onshore locations.

The objective of Volume 2, Advanced Piping Design, on the same subject, is to look into more detail at the design of process piping systems in specific locations around the various items of process equipment that would be typically found in a petrochemical or oil and gas processing facility.

For Volume 2, I enlisted the direction and support of Rutger Botermans, from Delft in The Netherlands, who is the author of this title. He wrote the text in a very direct style to avoid any misinterpretation. The bullet point/checklist format allows the reader so see quickly if he or she has considered the point when laying out the piping system. Rutger and his company, Red-Bag, have a great deal to offer the industry; and I look forward to working with him again on other projects.

Peter Alspaugh, from Bentley Systems Incorporated, was responsible for the CAD-generated drawings and worked under a very tight time frame over the Christmas period. During this particularly busy period, Peter also became a proud father; and I am sure that we all wish Alspaugh junior health and happiness.

Finally, my thanks go to Paul Bowers and Richard Beale, Canadian gentlemen who, at very short notice, also during Christmas 2007, carried out a peer reading exercise of the original manuscript. Both gentlemen are very experienced "pipers," and their comments were quite constructive and, where possible, incorporated into the book. At this very late stage, during the peer reading phase, it was not possible to include some of their ideas. They made some very interesting suggestions that would be ideally suited to a "stand-alone" book on the topics that they think need addressing on the subject of piping design. I look forward to the opportunity to work with them again.

Volume 2 is not intended to be the finale on the subject of piping design; and because of the nature of the subject, a practitioner never stops learning about the many facets of design, fabrication, erection, inspection, and the testing of piping systems.

A process facility is made up of numerous items of equipment, which are used to efficiently change the characteristics of the process flow and change the feedstock initially introduced into the facility. The result is a final product that can be dispatched for distribution to an end user for consumption or further refining.

For a process plant to operate effectively and efficiently and therefore maximize the commercial returns, it is very important that the size, routing, and the valving of process and utility piping systems are optimized to allow them to work efficiently.

As with Volume 1, we target the young and intermediate designers and engineers, individuals with a fundamental knowledge of the subject of piping systems who are interesting in expanding their set of technical skills to the next level.

This volume covers the various major types of equipment that make up an integrated process facility:

> Pumps, to move process liquids.
>
> Compressors, to move process gases.
>
> Heat exchangers, to transfer heat from one product to another.
>
> Fired heaters, for direct heating of a product.
>
> Tanks, to store process and utility fluids.
>
> Columns towers, for distillation of products.
>
> Relief systems, to protect the piping system from over pressurization.
>
> Pipe racks and tracks, to route and support the piping systems.

This second volume is not intended to be a conclusion on the subject of piping, and the reader is recommended to continue to read and research the subject in the future, because the subject is so diverse and suffers no shortage of opinions.

—Peter Smith
Leiden 2008

CHAPTER 1

Basic Plant Layout

1.1 Introduction

1.1.1 General

No two oil and gas processing facilities are exactly the same; however, they share similar types of process equipment, which perform specific functions. The following are significant items of equipment that are discussed in more detail in this book:

- Pumps for the transportation of liquids.
- Compressors for the transportation of compressible fluids.
- Exchangers for the transfer (exchange) of heat from a heating medium to a fluid.
- Fired heaters.
- Columns.
- Tanks for the storage of compressible and noncompressible fluids.
- Pipe racks and pipe ways for the routing of process and utility pipework between equipment.

To allow the facility to function safely and efficiently, to maximize its commercial profitability, and to result in the optimum layout, the interrelationships among the various types of process equipment must be carefully considered. As the layout is developed, compromises often must be made, and the preference generally is the safer option.

All operators of process plants have the same objectives, which is to produce a stable product that meets the end users specification and to maximize the commercial potential of the feedstock for the life of the plant. Even with this common goal operators have subtle differences in the way they have their facilities designed; therefore, the word *generally* is used liberally in these pages. *Generally* means that it is common practice, but it is not a mandatory requirement.

Listed next are the considerations that have to be reviewed when positioning the equipment during the development of the plant layout. They have not been listed in an order of priority; however, safety is listed at the top as the most significant issue.

- Safety: fire, explosion, spillage, escape routes for personnel, and access for firefighters.
- Process flow requirements that result in an efficient plant.
- Constructability.
- Segregation of areas for hazardous and nonhazardous materials.
- Operability and maintainability.
- Available plot area, geographical limitations.
- Relationship to adjacent units or other facilities within the plant.
- Economics.
- Future expandability.
- Security: control of access by unauthorized personnel.
- Meteorological information: climate, prevailing and significant wind direction.
- Seismic data.

Equipment should be laid out in a logical sequence to suit the process flow. Fluid flow requirements (for example, gravity flow systems, pump suction heads, and thermosyphonic systems) often dictate relative elevations and necessitate the need for structures to achieve the different elevations. Limitations of pressure or temperature drop in transfer lines decide the proximity of pumps, compressors, furnaces, reactors, exchangers, and the like.

Equipment piping should be arranged to provide specified access, headroom, and clearances for operation and maintenance. Provision should be made to minimize the disturbance to piping when dismantling or removing equipment (for example, without removing block valves), including the use of and free access for mobile lifting equipment. Pumps should be located in rows adjacent to their pipe ways and near the equipment from which they take suction. The top nozzles of pumps should be located in the vicinity of overhead steel, such as a beam at the side of the pipe rack, to facilitate piping support.

Plant layout requires input from the following discipline engineers;

- HSE (health, safety, and environment).
- Process.
- Piping.
- Mechanical (rotating and vessels).
- Civil and structural.
- Instrumentation.
- Electrical.

Once the relevant information has been sourced, several meetings probably will take place between engineers of these disciplines to develop a plant layout that will satisfy the project's requirements.

As mentioned previously, no two operating companies have exactly the same philosophies; however, they share the same basic common objective, which is to design, construct, and operate a plant that is both safe and economic for the duration of the facility's operating life.

The following lists of points should be considered when evaluating the layout of a plant and the relationships among the various items of equipment. They are not necessarily mandatory and could be changed, based on aesthetics, economics, safety, maintenance, or the operator's experience.

1.1.2 Pumps

- Locate pumps close to the equipment from which they take suction. This is an important consideration.
- Consideration should be made to locate pumps under structures or with their motor ends under a pipe rack, allowing an access aisle for mobile handling equipment.
- Pump suction lines generally are larger than discharge lines, to avoid problems arising from a low net positive suction head (NPSH).
- End suction with top discharge is the preferable option for pumps, when taking suction directly from tanks or vessels located at grade.
- Pumps should be arranged in rows with the center line of discharges on a common line.
- Clearances between pumps or pumps and piping generally are a minimum of 900 mm.

1.1.3   Compressors

- It is important to locate reciprocating compressors, anchors, and restraints for pipes belonging to the compressor system on foundations that are independent of any building, structure, or pipe track or rack. This independence gives the associated piping stability and minimizes unnecessary fatigue and possible failure.
- Spacing between compressors and other equipment varies with the type of machine and its duty.
- Particular attention must be paid to withdrawal of engine and compressor pistons, cam shaft, crank shaft, and lube oil cooler bundle; cylinder valve maintenance clearance with the least possible obstruction from piping supports.
- Compressors generally are provided a degree of shelter, that is, a sheets building. Keep the sides up to 8 feet above grade and open and vent the ridge to allow for escape of flammable gas, which might leak from the machines.
- Certain types of compressors, owing to the height of the mass foundation above grade level, require a mezzanine floor of a grid construction to avoid trapping any gas and for operation and maintenance.

1.1.4   Exchangers

- Tubular exchangers usually have standard length tubes of 2.5, 4, 5, and 6 m.
- Whenever possible locate exchangers at grade to facilitate maintenance and tube withdrawal.
- Two or more shells forming one unit can be stacked or otherwise arranged as indicated on the exchanger specification sheet, which is delineated by the manufacturer.
- Exchangers with dissimilar service can be stacked, but rarely more than three high, except for fin-tube-type units.
- Horizontal clearance of at least 900 mm should be left between exchangers or between exchangers and piping.
- Where space is limited, clearance may be reduced between alternate exchangers, providing sufficient space is left for maintenance and inspection access.
- Tube bundle removal distance should be a minimum of a tube length plus 900 mm. Minimum removal distance plus 600 mm should be left behind the rear shell cover of floating head exchangers.

- Where a rear shell cover is provided with a davit, allow clearance for the full swing of the head. Set overhead vapor exchangers or condensers at such elevation that the exchanger is self-draining.
- Arrange outlets to a liquid hold pot or trap, so that the underside of the exchanger tubes is above the liquid level in the trap.
- Arrange exchangers so that the fixed end is at the channel end.
- Vertical exchangers should be set to allow lifting or lowering of the tube bundle.
- Consult the Vessel Department as to the feasibility of supporting vertical exchangers from associated towers.
- Space should be left free for tube or bundle withdrawal, with the exchanger channels preferably pointing toward an access area or road.
- If an exchanger is situated well within the plot, leave a free area and approach for mobile lifting equipment.
- Air fin exchangers, preferably, should be located in a separate row outside the main equipment row, remote from the central pipe way.
- Consider locating air fin exchangers over the central pipe way if plot space if very limited.

1.1.5 Fired Heaters

- Fired heaters should be located at least 15 m away from other equipment that could be a source of liquid spillage or gas leakage.
- To avoid accumulation of flammable liquids, no pits or trenches should be permitted to extend under furnaces or any fired equipment, and if possible, they are to be avoided in furnace areas.
- Ensure ample room at the firing front of the fired heater for operation and removal of the burners and for the burner control panel, if required.
- Bottom-floor fired furnaces require adequate headroom underneath the furnace. Wall fired furnaces require an adequate platform width with escape routes at each end of the furnace.
- Apart from an adequate platform and access to the firing front, other structural attachments and platforms around

furnaces should be kept to a minimum. Peepholes should be provided only where absolutely necessary. Access by means of a stepladder is sufficient.

- Arrange fired heaters on a common center line, wherever possible.

- Provide unobstructed space for withdrawal.

- Operation and maintenance platforms should be wide enough to permit a 1-m clear walkway.

- Escape ladders should be provided on large heaters.

- Vertical heaters usually are supplied with stub supporting feet; ensure drawings show adequate supports elevated to the required height.

- Headroom elevation from the floor level to the underside of heater should be 2.3 m, to provide good firing control operation.

1.1.6   Columns

- Columns usually are self-supporting with no external structures.

- Circular or segmental platforms with ladders are supported from the shell.

- The maximum allowable straight run of a ladder before a break platform should not exceed 9 m.

- The factors influencing column elevation are the provision of a gravity flow system and installation of thermosyphon reboilers.

- Depending on the plant arrangement, columns may have to be elevated to a height in excess of the normal requirements to allow for headroom clearance from lower-level piping off-takes.

- The skirt height of all columns or vessels providing suction to pumps, particularly if handling hot or boiling liquids, should be adequate for the pump NPSH requirements.

- Access platforms should be provided on columns for all valves 3" and above, instrument controllers and transmitters, relief valves, manholes and blinds or spades, and other components that require periodic attention.

- For access to valves 2" and smaller, indicating instruments, and the like, a ladder is acceptable.

- Platforms for access to level gauges and controllers should not be provided if underside of supporting steelwork is less than normal headroom clearance from grade.
- Adjacent columns should be checked, so that platforms do not overlap. For layout, 2.0–2.5 m between shells, depending on insulation, should suffice.
- Allow a 900 mm minimum clearance between column foundation and the adjacent plinth.
- Provide clearance for the removal of internal parts and attachments and for davits at top of columns, if relevant.
- The center line of manholes should be 900 mm above any platform.
- Horizontal vessels should be located at grade, with the longitudinal axis at a right angle to the pipe way, if possible.
- Consider saving plot space by changing vessels from the horizontal to the vertical, if possible, and combining vessels together with an internal head (subject to project or process approval).
- The size and number of access platforms on horizontal vessels should be kept to a minimum and are not to be provided on horizontal vessels or drums when the top of the vessel is 2.5 m or less from the grade.
- The channel end of vessels provided with internal tubular heaters should face toward an open space. The withdrawal area must be indicated on studies, general arrangements (GAs), and plot plans.
- Internal agitators or mixers are to be provided with adequate clearance for removal. Removal area must be indicated on studies, GAs, and plot plans

1.1.7 Tanks

- The layout of tanks, as distinct from their spacing, should always take into consideration the accessibility needed for firefighting and the potential value of a storage tank farm in providing a buffer area between process plant and, for example, public roads and houses, for safety and environmental reasons.
- The location of tankage relative to process units must be such as to ensure maximum safety from possible incidents.

1.1.8   Pipe Racks and Pipe Ways

- Ideally, all piping within a process area should be run above grade; however, for many reasons this is not possible. Trenched or buried piping should be avoided but, sometimes, is unavoidable. Pipe racks at higher elevations, using supports, are preferred.

- Pipe racks may contain one, two, or more layers of pipework; however, triple-layer pipe racks should be limited to very short runs.

- Run piping external to the process area at grade on sleepers generally 300 mm high. Pipe ways at grade are cheaper but more liable to interfere with access.

- Locate the largest bore and the heaviest piping as close to stanchions as possible.

- Lines requiring a constant fall (relief headers) can be run on cantilevers from pipe-rack stanchions or on vertical extensions to pipe-track stanchions.

- Run the hot line requiring expansion loops on the outside edge of pipe way to permit loops to have greatest width over the pipe way and facilitate nesting.

- Takeoff elevations from pipe ways should be at a constant elevation, consistent with the range of pipe sizes involved.

- Change elevation whenever banks of pipes, either on pipe ways at grade or at higher elevations on pipe racks, change direction.

- Elevations to the underside of pipe racks should be the minimum for operation and mobile maintenance equipment and consistent with allowable clearances.

- Open pipe trenches may be used between plants where there is no risk of flammable vapors collecting.

- It sometimes is convenient to run open trenches alongside roadways. (Soil from the trench can be used to build up the road.)

- Where a pipe way or road changes from a parallel direction, the pipe generally is run beneath the road.

- Occasionally, it is permissible to run pipes in trenches to overcome a difficult piping problem. Such trenches should be of concrete, drained, and covered.

- Although trenched piping is to be avoided, due to the expense and hazards associated with open trenches, piping buried underground is acceptable, provided the pipe is adequately protected and below the frost line.

- The sizing and arrangement of underground piping should be fixed early to ensure that installation is simultaneous with foundation work. (Many drains, sewers, and cableways, which do not require attention, are run underground below the frost line.)

- Leave space for draw boxes on cableways, anchors on underground cooling water pipes, and manholes on sewers. Fire mains should be located between the perimeter road and the plant.

1.2   Guidelines for Laying out the Plant

1.2.1   General

All elevations in Tables 1–1 through 1–4 refer to a nominal 100 m. The point 100 elevation is taken as the high point of paving in the paved areas. This should be common throughout the plant. Equipment elevations referring to grade elevations of 100 m are as shown in Table 1–2.

Table 1–1   Access Clearances

Description	Minimum Clearance (m)		
	Headroom	Width	Other
Primary access roads (carrying major equipment)	6	6	10.5, inside corner radius
Secondary roads	5.1	4.8	4.5 inside corner radius
Minor access roads	5.1	3.6	—
Yard piping	3	—	—
Platform, walkways, passageways, working areas, stairways	2.1	1 working platforms	—
Clearance from face of manhole	2.1	1	Manhole center approx. 1 m above platform
Railways	To suit local codes	—	—

Table 1–2   Elevation

Description	Minimum (m)
Open-air paved area high point of paving	Construction to determine
Underside of baseplates for structural steel	Construction to determine
Stair and ladders pads	Construction to determine
Underside of baseplates vessel and column plinths	Construction to determine
Top of pump plinths	Construction to determine

Table 1–3   Valve Access

Item	Size	Minimum Access From		
		Fixed Ladders	Edge of Platform	Location over Platform or Grade
Exchanger heads	All	—	—	X
Operational valves	2 and under	X	—	—
Operational valves	Over 2	—	X	—
Motor operated valves	All	—	X	—
Control valves	All	—	—	X
Relief valves (process)	2 and over	—	—	X
Block valves		Accessible by portable ladder		
Battery limit valves etc.	All	Edge of platform access where client's specifically requests; otherwise, no access		
Pressure instruments	All	X	—	—
Temp. instruments	All	X	—	—
Sample points	All	—	X	—
Try cocks	All	X	—	—
Gauge glasses	All	X	—	—

Table 1–3 Valve Access (cont'd)

Item	Size	Minimum Access From		Location over Platform or Grade
		Fixed Ladders	Edge of Platform	
Level controllers	All	—	X	—
Process blinds and spades	All	—	—	X
Walkways	All	—	—	X
Handholes	All	—	X	—
Nozzles	All	No access provided		
Vessel vents	All	X	—	—
Line drains and vents		No access provided		

Table 1–4 Maintenance Facilities

Equipment	Part Handled	Handling Facilities
Reactors, vessels and columns.	Manhole covers	Davits or hinges for swinging open
	Internal, requiring regular removal or servicing	Trolley beams or davits for lowering from holes to grade
	Fixed bed reactors, catalyst change, etc.	Provided as specially specified to enable catalyst to be off-loaded and loaded
Floating head exchangers	Tube bundles	All such provided with jackbolts to break joints; bundles assumed to be handled by mobile equipment
	Exchanger heads, channel cover, bonnets	No special provision
Vertical exchangers	Removable tube bundles	Overhead trolley beam or davit

Table 1–4   Maintenance Facilities (cont'd)

Equipment	Part Handled	Handling Facilities
Pump	Any part	None
Centrifugal compressors	Rotating parts	Overhead trolley beams or cranes
Piping	Relief valves, 2 nominal bore and larger	Hitching point or davit for lowering to grade
	Blanks, blank flanges, and swing elbow weighing more than 300 lbs (125 kg)	Overhead hitching point or davit only when subject to frequent removal for maintenance

Ascertain soil-loading considerations and site contours before fixing the final layout. Considerable variations occur in allowable soil loads throughout site areas. It may be advantageous to locate heavy equipment in the best soil-loading area. Use existing contours, so that the quantity of earth movement due to cut and fill may be reduced substantially by intelligent positioning of the equipment.

1.2.2   Safety

- Provide a sufficient clear area between critical or high-temperature items of equipment. Clear routes for operators with two or more escape ladders or exits at extremities.
- Clear routes for access by firefighting equipment.
- Do not allow areas classified as hazardous to overlap the plot limits or extend over railways where open firebox engines are likely to be employed.
- Stacks should be located so that prevailing winds do not blow smoke over the plant. Try not to locate the plant where it will receive dust, smoke, spray, or effluent from a neighboring plant.
- Avoid using locations polluted by continuous drift of dust, smoke, and the like.
- If the plant is to be located in an existing refinery or factory site, line up with existing roads, columns, and stacks.
- The location of external railways, pipe ways, cableways, sewers and drains, and so forth, also may influence the final orientation of the plant.

- When railway facilities are required, avoid boxing in the plant by branch lines.

- Hazardous areas from other existing plants or equipment may extend over the plant limit. This could effectively reduce plot size and thus influence the plant layout philosophy.

1.2.3 Hazardous and Toxic Areas

Equipment items considered a possible source of hazard should be grouped and located separately, if possible and economic. Examples are furnaces, flare stacks, or other direct-fired equipment containing an open flame and rotating or mechanical equipment handling flammable or volatile liquids that could easily leak or spill. Equipment handling acids or other toxic materials that could cause damage or danger by spillage should be grouped and contained within a bunded area.

Location of Control Rooms

Locate control rooms 15 m or more from equipment that, in operation or during maintenance, can create a hazard. (If not practicable, pressurize.) Ensure the maximum length of a cable run to any instrument is no more than 90 m.

Location of Buildings

- Locate, for example, offices, first-aid rooms, cafeterias, garages, fire station, warehouses, gas holders, and workshops, at a minimum of 30 m from any hazard.

- Unpressurized substations and switchrooms should be a minimum of 15 m from any hazard.

- The determination of dangerous areas and their safety requirements should be in accordance with the Institute of Petroleum Safety Codes or, where this is not recognized, the applicable national code(s).

- Local bylaws and fire codes, whose requirements may be more stringent or specific than the preceding codes, take precedence.

1.2.4 Constructability, Access, and Maintenance

- The overall plant arrangement must be reviewed for constructability, operation, safety, and maintenance. Large items of equipment or towers that require special lifting gear need adequate access to lift these into place.

- Large equipment positioned close to boundary limits may require erection from the outside.

- Ascertain whether sufficient space will be available at the construction phase.

- Operation and maintenance should be reviewed by the eventual operating company. Give consideration to maintenance access to air fins and the like above pipe tracks.

- Consider the location of equipment requiring frequent attendance by operating personnel and the relative position of the control room to obtain shortest, most direct routes for operators when on routine operation.

Clearances

As in Table 1–1, access clearances between adjacent plants should at least equal those for primary access roads. The space between edge of any road and nearest equipment must be no less than 1.5 m.

- Adequate road access with properly formed roads must be provided for known maintenance purposes; for example, the compressor house, large machinery areas, reactors, or converters requiring catalyst removal and replacement.

- Equipment requiring infrequent maintenance, such as exchanger tube bundles and tower internals, need adequate level clear space for access and removal purposes.

- The ground need not be specifically built up to take loads other than a surfacing of granite chips or similar, as duckboards, gratings, or other temporary material can be laid at the time when the plant is under maintenance.

Paving

- Within the process area, minimal concrete paving should be supplied for walkways interconnecting major items of equipment, platforms, stairways, and buildings.

- Paving should be supplied around pumps or other machinery located in the open, underneath furnaces, and any other areas where spillage is likely to occur during normal operation.

- Areas containing alkalis, acids, or other chemicals or toxic materials should be paved and bunded to prevent spillage spreading. Other areas of the plant are to be graded and surfaced with granite chips or similar material.

Insulation

- Insulation may be applied to vessel supports or stanchions of structures for fire protection, thus decreasing the available free space for access and siting of pipework, instruments, or electrical equipment.

- In particular, note the thickness of insulation of very high- or low-temperature piping, which may considerably increase the effective outside dimension of pipe to be routed.

- For low-temperature insulation, additional clearance must be provided around control valves, instrumentation, and the like. Consider the additional weight of insulation and reduced centers of supports necessary to support heavily insulated pipe.

Instrumentation

- All operating valves 3" and larger are to be accessible either from grade or a suitable platform with a maximum 2.0 m above working level to center of handwheel.

- Small operating valves can be reached from a ladder. Valves installed for maintenance and shutdown purposes (other than operating) can be reached by portable ladder.

- Otherwise, extension spindles or suitable remote operating gear should be provided but not on valves 1½" and smaller. The minimum access to be provided is as shown in Table 1–3.

Relief Valve Systems

Closed relief valve systems should be arranged to be self-draining and should not contain pockets where liquids may condense and collect to provide any back pressure.

1.2.5 Economics

- Apart from process restrictions, position the equipment for maximum economy of pipework and supporting steel. As compact a layout as possible with all equipment at grade is the first objective, consistent with standard clearances, construction, and safety requirements.

- Minimize runs of alloy pipework and large-bore pipe without the introduction of expensive expansion devices.

- Optimize the use of supporting structures in concrete or steel by duplicating their application to more than one item of equipment and ensuring that access ways, platforms, and the like have more than one function.

- Space can be saved by locating equipment over the pipe rack. Pumps in general should be located with their motors underneath the main pipe rack.

1.2.6   Aesthetics

Attention should be paid to the general appearance of the plant. An attractively laid-out plant with equipment in straight lines usually is economical.

- Preference should be given to use of a single, central pipe way with a minimum number of side branches and equipment laid out in rows on either side.
- Buildings, structures, and groups of equipment should form a neat, symmetrical, balanced layout, consistent with keeping pipe runs to a minimum.
- Arrange towers and large vertical vessels in rows with a common center line if of similar size but lined up with a common face if diameters vary greatly. If adjacent to a structure, the common face should be on the structure side.
- The center lines of exchanger channel nozzles and pump discharge nozzles should be lined up.
- Piping around pumps, exchangers, and similar ground-level equipment should be run at set elevations, one for north-south and another for east-west elevations wherever possible (similarly, racked pipework). These elevations being to the bottom of the pipe or the underside of the shoe for insulated lines. This also should help achieve a common elevation for off-takes from pipe ways.
- If possible, duplicated streams should be made identical.
- Where possible, handed arrangements should be the second choice.
- Follow this principle for this similar equipment sequences within the process stream; for example, a fractionators tower with overhead condensers, reflux drum pumps, and a reboiler is a system that could be repeated almost identically for different towers having a different process duty. The advantages are design and construction economy, improved maintenance, and operating efficiency.

1.2.7 Instruments to Assist Initial Layout

Table 1–5 lists some likely devices, the probable number fitted to various types of equipment, and the design points affected. In-line instrument elements—flow elements (orifice, plates, venturi, turbine, pressure differential, etc.), control valves (globe, butterfly, ball, etc.), relief valves, thermowells—are listed in Table 1–6.

Table 1–5 Devices Fitted to Equipment

Equipment	Devices and Probable Number of Items Fitted
Distillation tower	PSV (pressure safety valve), 1
	PIC (pressure indicating controller), 1
	FRC (flow recording controller), 3
	TR (temperature recorder), multipoint, 6 channel, 1
	TI (temperature indicator), 6
	PI (pressure indicator), 6
	Analyzer (single stream), 1
	LG (level gauge), 2
	LI (level indicator), 1
	LIC (level indicating controller), 1
Reflux drum surge drum buffer storage feed tank product tank	LG (level gauge), 3
	LIT (level indicating transmitter), 1
	PI (pressure indicator), 1
	TI (temperature indicator), 1
	PSV (pressure safety valve), 1
	PIC (pressure indicating controller), 1
Reactor	PI (pressure indicator), 6
	TI (temperature indicator), 6
	PSV (pressure safety valve), 1
	TR (temperature recorder), multipoint, 50 channels, 1

Table 1–5   Devices Fitted to Equipment (cont'd)

Equipment	Devices and Probable Number of Items Fitted
	FIC (flow indicating controller), 2
	LIC (level indicating controller), 1
	Analyzer, 1
	PIC (pressure indicating controller), 1
	TIC (temperature indicating controller), 1
Compressor (axial flow)	PI (pressure indicator), 4
	DPC (differential pressure controller), 1
	PIC (pressure indicating controller), 1
	FR (flow recorder), 1
	TI (temperature indicator), multipoint, 12 channels, 1
	Vibration indicator, 2
	NRV, programmer and logic system, damped to prevent reverse flow, 1
	Shutdown system, 1
Compressor driver (steam turbine)	PI (pressure indicator), 4
	FRC (flow recording controller), 1
	TI (temperature indicator), 4
	Shutdown valve, 1
Exchanger	TRC (temperature recording controller), 1
	TI (temperature indicator), 6
	PI (pressure indicator), 2
	LG (level gauge), 1
	PSV (pressure safety valve), 1
Furnaces	FRC (flow recording controller), 4

Table 1–5   Devices Fitted to Equipment (cont'd)

Equipment	Devices and Probable Number of Items Fitted
	TRC (temperature recording controller), 1
	PIC (pressure indicating controller), 1
	Flame detector (2)
	Local panel (1)
	PI (pressure indicator), 12
	TI (temperature indicator), 6
	Multichannel temperature indicator, 1
	O_2 analyzer, only where BFW or steam is circulating, 1
	pH analyzer, 1
	Conductivity monitor
	LG (level gauge), 3
	LIC (level indicating controller), 1
	PSV (pressure safety valve), 3
	PCV (pressure control valve), 3

Table 1–6   In-Line Instrument Elements

Device	Type	Design Points Affected
Flow		
Fe 1	Pipe section with sensing element	Flange rating, size, overall length, orientation
Fe 2	Pitot tube	Location, straight length, connection size and type
Fe 3	Orifice, nozzle, venture tube	Location, straight length, orientation; flange size and rating; position, size, type of instrument tappings
Fe 4	Elbow tube	Size, end connections, orientation, straight length

Table 1–6   In-Line Instrument Elements (cont'd)

Device	Type	Design Points Affected
Fe 5	Target meter transmitter	Orientation, straight length; flange rating, connections, face-to-face insertion
Fe 6	Vortex meter	Orientation, straight length, flange rating, face-to-face insertion
Fe 7	Hot wire	Consult instrumentation specialist
Fe 8	Variable area meter	Vertical only, upward flow only; orientation of connections, sizes, and type
Fe 9	Magnetic flowmeter	Overall length, size, connections vertical or horizontal, no straight lengths
Fe 10	Turbine meter	Straight length, with or without pipe section, usually horizontal end connection and size (common to use upstream filter and sometimes degassing)
Fe 11	Positive displacement	Orientation one way only, weight, no straight lengths; connecting as per vendor literature
Fe 12	Sonic flowmeter	Consult instrumentation specialist
Fe 13	Weight rate	Consult instrumentation specialist
Fe 14	Radioactive	Consult instrumentation specialist
Fe 15	Photo electric	Consult instrumentation specialist
Fe 16	Channels and flumes	Mostly civil engineering
Fe 17	Vane type	Spool piece = face-to-face end connections
Temperature		
TE 1	Thermocouple	
TE 2	Resistance bulb	Location, increase in pipe diameter and elbows; connection size and type
TE 3	Filled system	
TE 4	Thermistor	
TE 5	Radiation	Location of window, heat protection

Table 1–6   In-Line Instrument Elements (cont'd)

Device	Type	Design Points Affected
Flame Failure		
Be	Photo electrical, color	Location of instrument and window
Analyzer		
An	Diverse methods including specific gravity and density	Usually with bypass line to drain on back to process, only occasionally in line; sometimes coaxial spool piece; face to face, flanges
Level Measurement		
An	Capacitance	Similar to temperature Te 1
Probe	Conductivity	Sometimes coaxial in spool piece
Proximity Switch		
	Feromagnetic)	Nonintrusive, location and mounting
	Magnetic	Nonintrusive
	Inductive	Instrusive, type Te 1
Pressure		
Differential pressure	Bourdon tubes capsules strain gauge	Small tapping, location, connections, size and type
Gauge Glass Level		
Level gauge	All types	Vertical only, nozzle spacing, connections
Interface Level (Gauge-Glass)		
Liquid gas	All types	Vertical only, nozzle spacing critical, connections
Speed Measurement		
	Magnetic	Consult instrumentation specialist

Table 1–6   In-Line Instrument Elements (cont'd)

Device	Type	Design Points Affected
	Strobe tachiometer	Consult instrumentation specialist
Valves		
PV, FV, TV, etc., PVC	Operation: electrical, hydraulic, pneumatic, self-operated	Nominal body size determined by flow criteria; face to face, connection sizes, flange rating often 300 lbs minimum as a standard; axis of movement of top works must be vertical, all other orientation prohibited; face-to-face dimensions do not always conform to industry standards
Safety Valves		
PSV	Spring opposed pressure	Free or closed venting; multiple valve relief; gauge valves with single operation of changeover, single isolation valves prohibited, minimum nozzle size laid down in the codes; Some inlet-outlet flange combinations excluded in standard manufacture, depending on application

CHAPTER 2

Pumps

2.1 Introduction

A pump is a machine used to generate a differential pressure to propel a liquid through a piping system from one location to another. Efficient transference of liquids, from equipment to equipment through various elevations, is essential for a process plant to function.

This chapter briefly discusses the types of pumps likely to be encountered in a process plant and highlights the special piping issues to be considered when piping up specific types of equipment.

It is suggested that the individual responsible for piping up a pump make himself or herself familiar with the inner workings of that particular pump, to be aware of the requirements of the primary piping (transporting the process liquid) and the ancillary piping (used for cooling, lubrication, drains, and vents to support the operation of the pump).

2.2 Types of Pumps

The three basic types of pump are centrifugal, reciprocating, and rotary (see Figure 2–1).

2.2.1 Centrifugal

Centrifugal pumps are the type most commonly used in an oil and gas processing facility; and because they have an industrial purpose, they are considered to be heavy duty.

Centrifugal pumps generally are more economic in service and require less maintenance than the other two options. The rotation of the impeller blades produces a reduction in pressure at the center of the impeller. This causes the liquid to flow onto the impeller from the

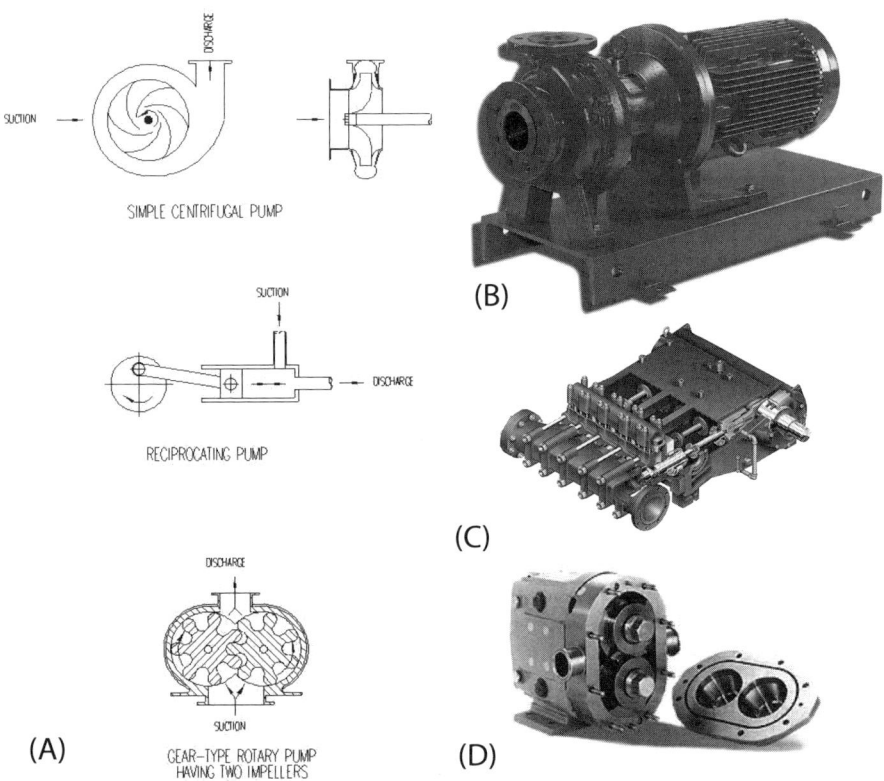

Figure 2–1   *(A) Pump types: (B) centrifugal pump, (C) multiplunger reciprocating pump, (D) rotary pump (courtesy of Red Bag/Bentley Systems, Inc., and 2.1B, BHP Pumps; 2.1C, Flowserve; 2.1D, Waukesha).*

suction nozzle. The fluid is thrown outward along the blades by a centrifugal force. The liquid then leaves the blade tips via the pump volute and finally leaves the pump through the discharge nozzle. This transference of the liquid is completed in a smooth, nonpulsating flow.

The three basic types of centrifugal pumps are

- A horizontal drive shaft with the pump drive mounted remote from the in-line piping.
- A vertical drive shaft with pump and drive mounted in-line with the piping.
- A vertical barrel type with direct immersion suction facility.

In each case, the type refers to drive shaft orientation.

The most common is the horizontal drive with its shaft in the horizontal plane. Vertical in-line pumps have their shafts vertical and the pump is installed in the piping system the same way as for an in-line valve. Vertical barrel pumps usually are single stage, but horizontal and vertical drive shaft types can be of single-, double-, or multistage design, depending on the design conditions (pressure and temperature) of the process fluid, type of fluid (noncorrosive, mildly corrosive, or highly corrosive), net positive suction head available, and to obtain the desired delivery pressures.

2.2.2 Reciprocating

Reciprocating pumps are used where a precise amount of fluid is required to be delivered, also where the delivery pressure required is higher than can be achieved with other types.

The fluid is moved by the means of a piston that travels in a cylinder. After being drawn into the cylinder through an inlet valve, the piston continues moving down the cylinder. As the piston moves back up the cylinder, the liquid is discharged at a preset pressure, controlled by a delivery valve.

The liquid is ejected from the cylinder into the piping system in pulses, which are transmitted to the suction and discharge piping; therefore, hold-down supports could be required on the piping system on the suction and the discharge side of the pump.

The three classes of reciprocating pumps are piston, plunger, and diaphragm. Piston pumps generally are used where medium to high delivery pressures are required, such as for a high-pressure flushing of vessel interiors and tanks. These can be obtained in multicylinder form and can be single or double acting. Plunger pumps usually are used for metering or proportioning a fluid. Frequently, a variable speed drive or stroke adjusting mechanism is provided to vary the flow as desired. Diaphragm pumps are invariably air driven and very compact, also no seals or packing is exposed to the liquid being pumped, which makes them ideal for handling hazardous or toxic liquids. These often are used for sump pump out.

2.2.3 Rotary

Rotary pumps are used to move heavy or very viscous fluids. These employ mechanical means such as gear, cam, and screw, to move the fluid. The two main classes of rotary pumps are gear and screw. Gear pumps usually are employed to pump oils and nonabrasive fluids. Screw pumps usually are used to pump heavy viscous fluids and nonabrasive slurries or sludge. Apart from maintaining good access to pumps for operation and maintenance, each case should be treated on an individual basis.

2.3   Types of Drivers

The three most common types of driver are the electric motor, diesel engine, and steam turbine. Each option has advantages and disadvantages, and selection is based on the application, location, availability of fuel or power, safety, and economics.

2.3.1   Electric

Electric motors are the most common pump driver and are of the totally enclosed, flameproof type suitable for zone 1 use. Their sizes range from small to very large, which require their own cooling systems.

2.3.2   Diesel

Diesel engines usually are to be found as drivers for fire water pumps, which are housed in a separate building away from the main complex.

2.3.3   Steam

Steam turbines used for pump drivers usually are single stage, and the pump that they drive is invariably for standby service (spare).

2.3.4   Gas Turbine

Gas turbines are considered if a local source of fuel is a readily available.

2.4   Applicable International Codes

Numerous international codes and standards apply to the various types of pump that could be used in hydrocarbon processing plants, and listed next are several of the most important ones, along with the scope of the document and associated table of contents.

The design and specifying of these specialized items of equipment are the responsibility of the mechanical engineering group; however, a piping engineer or designer benefits from being aware of these documents and reviewing the appropriate sections that relate directly to piping or a mechanical-piping interface:

- ANSI/API Standard 610.
- API Standard 613.

- API Standard 614.
- API Standard 670.
- API Standard 674.
- API Standard 675.
- API Standard 676.
- API Standard 677.
- ANSI/API Standard 682.
- API Standard 685.

2.4.1   ANSI/API Standard 610 and ISO 13709:2003 (Identical). Centrifugal Pumps for Petroleum, Petrochemical and Natural Gas Industries

Scope

This international standard specifies requirements for centrifugal pumps, including pumps running in reverse as hydraulic power recovery turbines, for use in petroleum, petrochemical, and gas industry process services.

The international standard is applicable to overhung pumps, between-bearings pumps, and vertically suspended pumps. Clause 8 provides requirements applicable to specific types of pump.

All other clauses of the international standard are applicable to all pump types. Illustrations are provided of specific pump types and the designations assigned to each one. This international standard is not applicable to sealless pumps.

Table of Contents

Introduction.
1. Scope.
2. Normative References.
3. Terms and Definitions.
4. Classification and Designation.
    4.1. General.
    4.2. Pump Designations.
    4.3. Units and Governing Requirements.
5. Basic design.
    5.1. General.
    5.2. Pump Types.
    5.3. Pressure Casings.

Annex E (Informative). Inspector's Checklist.

Annex F (Normative). Criteria for Piping Design.

Annex G (Informative). Materials Class Selection Guidance.

Annex H (Normative). Materials and Material Specifications for Pump Parts.

Annex I (Normative). Lateral Analysis.

Annex J (Normative). Determination of Residual Unbalance.

Annex K (Normative). Seal Chamber Runout Illustrations.

Annex L (Informative). Vendor Drawing and Data Requirements.

Annex M (Informative). Test Data Summary.

Annex N (Informative). Pump Datasheets.

Bibliography.

2.4.2 API Standard 613, Special Purpose Gear Units for Petroleum, Chemical and Gas Industry Services

Scope

This standard covers the minimum requirements for special-purpose, enclosed, precision single- and double-helical one- and two-stage speed increasers and reducers of parallel shaft design for petroleum, chemical, and gas industry services. This standard is intended primarily for gear units in continuous service without installed spare equipment. Gear sets furnished to this standard should be considered matched sets.

Table of Contents

1. General.
    1.1. Scope.
    1.2. Applications.
    1.3. Alternative Designs.
    1.4. Conflicting Requirements.
    1.5. Definition of Terms.
    1.6. Reference Publications.
    1.7. Standards.
    1.8. Units of Measure.
2. Basic Design.
    2.1. General.
    2.2. Rating.

Appendix J. Rating Comparison API 613 vs. AGMA 2101.

Appendix K. Shaft End Sizing Method.

Appendix L. Typical Mounting Plates.

2.4.3 API Standard 614. Lubrication, Shaft-Sealing, and Control-Oil Systems and Auxiliaries for Petroleum, Chemical and Gas Industry Services

Scope

This international standard covers the minimum requirements for lubrication systems, oil-type shaft-sealing systems, dry gas face-type shaft-sealing systems, and control-oil systems for general- or special-purpose applications.

General-purpose applications are limited to lubrication systems. These systems may serve equipment such as compressors, gears, pumps, and drivers. This standard does not apply to internal combustion engines.

This international standard is intended to be used for services in the petroleum, chemical, and gas industries as well as other industries by agreement. This standard is separated into four distinct chapters. Chapters 2, 3, and 4 are to be used separately in conjunction with Chapter 1.

Table of Contents

1. Scope.
2. Referenced Publications.
3. Definition of Terms.
4. General.
   4.1. Alternative Designs.
   4.2. Conflicting Requirements.
   4.3. System Selection.
5. Piping.
   5.1. General.
   5.2. Oil Piping.
   5.3. Instrument Piping.
   5.4. Process Piping.
   5.5. Intercoolers and Aftercoolers.
6. Instrumentation, Control, and Electrical Systems.
   6.1. General.

2.4.4   API Standard 670. Machinery Protection Systems

Scope

This standard covers the minimum requirements for a machinery protection system measuring radial shaft vibration, casing vibration, shaft axial position, shaft rotational speed, piston rod drop, phase reference, overspeed, and critical machinery temperatures (such as bearing metal and motor windings). It covers requirements for hardware (transducer and monitor systems), installation, documentation, and testing.

Table of Contents

1.   General.

1.1.   Scope.

    1.2. Alternative Designs.

    1.3. Conflicting Requirements.

2. References.

3. Definitions.

4. General Design Specifications.

    4.1. Component Temperature Ranges.

    4.2. Humidity.

    4.3. Shock.

    4.4. Chemical Resistance.

    4.5. Accuracy.

    4.6. Interchangeability.

    4.7. Scope of Supply and Responsibility.

5. Conventional Hardware.

    5.1. Radial Shaft Vibration, Axial Position, Phase Reference, Speed Sensing, and Piston Rod Drop Transducers.

    5.2. Accelerometer-Based Casing Transducers.

    5.3. Temperature Sensors.

    5.4. Monitor Systems.

    5.5. Wiring and Conduits.

    5.6. Grounding.

    5.7. Field-Installed Instruments.

6. Transducer and Sensor Arrangements.

    6.1. Location and Orientation.

    6.2. Mounting.

    6.3. Identification of Transducers and Temperature Sensors.

7. Inspection, Testing, and Preparation for Shipment.

    7.1. General.

    7.2. Inspection.

    7.3. Testing.

    7.4. Preparation for Shipment.

    7.5. Mechanical Running Test.

    7.6. Field Testing.

8. Vendor's Data.

    8.1. General.

    8.2. Proposals.

    8.3. Contract Data.

Appendix A. Machinery Protection System Data Sheets.

Appendix B. Typical Responsibility Matrix Worksheet.

Appendix C. Accelerometer Application Considerations.

Appendix D. Signal Cable.

Appendix E. Gearbox Casing Vibration Considerations.

Appendix F. Field Testing and Documentation Requirements.

Appendix G. Contract Drawing and Data Requirements.

Appendix H. Typical System Arrangement Plans.

Appendix I. Setpoint Multiplier Considerations.

Appendix J. Electronic Overspeed Detection System Considerations.

2.4.5 API Standard 674. Positive Displacement Pumps—Reciprocating

Scope

This standard covers the minimum requirements for reciprocating positive displacement pumps for use in service in the petroleum, chemical, and gas industries. Both direct-acting and power-frame types are included. See API Standard 675 for controlled-volume pumps and Standard 676 for rotary pumps.

Table of Contents

2.4.6   API Standard 675. Positive Displacement Pumps—Controlled Volume

Scope

This standard covers the minimum requirements for controlled-volume positive displacement pumps for use in service in the petroleum, chemical, and gas industries. Both packed-plunger and diaphragm types are included. Diaphragm pumps that use direct mechanical actuation are excluded. See API Standard 674 for reciprocating pumps and Standard 676 for rotary pumps.

Table of Contents

1. General.
    1.1. Scope.
    1.2. Alternative Designs.
    1.3. Conflicting Requirements.
    1.4. Definition of Terms.
    1.5. Referenced Publications.
    1.6. Unit Conversion.
2. Basic Design.
    2.1. General.
    2.2. Pressure-Containing Parts.
    2.3. Liquid End Connections.
    2.4. Pump Check Valves.
    2.5. Diaphragms.
    2.6. Packed Plungers.
    2.7. Relief Valve Application.
    2.8. Gears.
    2.9. Enclosure.
    2.10. Drive Bearings.
    2.11. Lubrication.
    2.12. Capacity Adjustment.
    2.13. Materials.
    2.14. Nameplates and Rotation Arrows.
    2.15. Quality.
3. Accessories.
    3.1. Drivers.
    3.2. Couplings and Guards.

2.4.7   API Standard 676 Positive Displacement Pumps—Rotary Manufacturing, Distribution and Marketing Department

Scope

This standard covers the minimum requirements for rotary positive displacement pumps for use in the petroleum, chemical, and gas industries. See API Standard 675 for controlled volume pumps and Standard 674 for reciprocating pumps.

Table of Contents

2.4.8 API Standard 677. General-Purpose Gear Units for Petroleum, Chemical and Gas Industry Services

Scope

This standard covers the minimum requirements for general-purpose, enclosed single- and multistage gear units incorporating parallel-shaft helical and right-angle spiral bevel gears for the petroleum, chemical, and gas industries. Gears manufactured according to this standard are limited to the following pitchline velocities: Helical gears should not exceed 60 m/s (12,000 ft/min), and spiral bevels should not exceed 40 m/s (8,000 ft/min). Spiral bevel gear sets should be considered matched sets.

Table of Contents

3.4. Controls and Instrumentation.

3.5. Piping and Appurtenances.

3.6. Special Tools.

4. Inspection, Testing, and Preparation for Shipment.

4.1. General.

4.2. Inspection.

4.3. Testing.

4.4. Preparation for Shipment.

5. Vendor's Data.

5.1. General.

5.2. Proposals.

5.3. Contract Data.

Appendix A. General-Purpose Gear Data Sheets.

Appendix B. Lateral Critical Speed Map and Mode Shapes for a Typical Rotor.

Appendix C. Typical Lube-Oil Systems.

Appendix D. Material Specifications for General-Purpose Gear Units.

Appendix E. Vendor Drawing and Data Requirements.

Appendix F. Referenced Specifications.

Appendix G. Spiral Bevel Gear-Tooth Contact Arrangement Requirements for Inspection.

Appendix H. Residual Unbalance Worksheets.

2.4.9 ANSI/API Standard 682 and ISO Standard 21049:2004 (Identical). Pumps—Shaft Sealing Systems for Centrifugal and Rotary Pumps

Scope

This international standard specifies requirements and gives recommendations for sealing systems for centrifugal and rotary pumps used in the petroleum, natural gas, and chemical industries. It is applicable mainly for hazardous, flammable, and toxic services, where a greater degree of reliability is required for the improvement of equipment availability and the reduction of both emissions to the atmosphere and life-cycle sealing costs. It covers seals for pump shaft diameters from 20 mm (0.75) to 110 mm (4.3).

This international standard also is applicable to seal spare parts and can be referred to for the upgrading of existing equipment. A classification system for the seal configurations covered by this international standard into categories, types, arrangements, and orientations is provided

2.4.10 API Standard 685 Sealless Centrifugal Pumps for Petroleum, Heavy Duty Chemical, and Gas Industry Services

Scope

This standard covers the minimum requirements for sealless centrifugal pumps for use in petroleum, heavy duty chemical, and gas industry services. Refer to Appendix U for application information.

Single-stage pumps of two classifications, magnetic drive pumps and canned motor pumps, are covered by this standard. Sections 2 through 8 and 10 cover requirements applicable to both classifications. Section 9 is divided into two subsections and covers requirements unique to each classification.

For process services not exceeding any of the following limits, purchasers may wish to consider pumps that do not comply with API Standard 685:

Maximum discharge pressure, 1900 kPa (275 psig).

Maximum suction pressure, 500 kPa (75 psig).

Maximum pumping temperature, 150°C (300°F).

Maximum rotation speed, 3600 rpm.

Maximum rated total head, 120 m (400 ft).

Maximum impeller diameter, 300 mm (13).

Table of Contents

1. Scope.
2. Referenced Publications.
3. Definition of Terms.
4. General.
   4.1. Unit Responsibility.
   4.2. Nomenclature.
5. Requirements.
   5.1. Units of Measure.
   5.2. Statutory Requirements.

Appendix A. Referenced Publications and International Standards.

Appendix B. Sealless Pump Data Sheets.

Appendix C. Sealless Pump Nomenclature.

Appendix D. Circulation and Piping Schematics.

Appendix E. Instrumentation and Protective Systems.

Appendix F. Criteria for Piping Design.

Appendix G. Material Class Selection Guide.

Appendix H. Materials and Material Specifications for Centrifugal Pump Parts.

Appendix I. Magnet Materials for Magnetic Couplings.

Appendix J. Procedure for Determination of Residual Unbalance.

Appendix K. Pressure Temperature Profiles in the Recirculation Circuit.

Appendix L. Baseplate and Soleplate Grouting.

Appendix M. Standard Baseplate.

Appendix N. Inspector's Checklist.

Appendix O. Vendor Drawing and Data Requirements.

Appendix P. Purchaser's Checklist.

Appendix Q. Standard Electronic Data Exchange File Specification.

Appendix R. Metric to U.S. Units Conversion Factors.

Appendix S. Withdrawn Summary.

Appendix U. Application Information.

2.5 Piping—Specific Guidelines to Layout

Numerous issues must be considered when piping up a specific type of pump. The following are based on the guidelines from several operators, based on their in-service experience. The pump manufacturer also should be consulted to see if any special issues should be considered.

2.5.1 Horizontal Centrifugal Pump Piping

Suction Piping

Net Positive Suction Head

- Centrifugal pumps must have their suction lines flooded at all times.

- The suction piping has to be designed to avoid cavitation and prevent vapor entering the pump. Therefore, suction lines should fall continuously from a sufficient height from overhead source to the pump and be adequately vented to minimize the presence of vapor.

- The minimum vertical height required from source of the liquid to pump suction is called the *net positive suction head.* This is critical for efficient pump operation and must not be reduced. Vessel elevations often depend on the NPSH of its associated pump. See Figure 2–2.

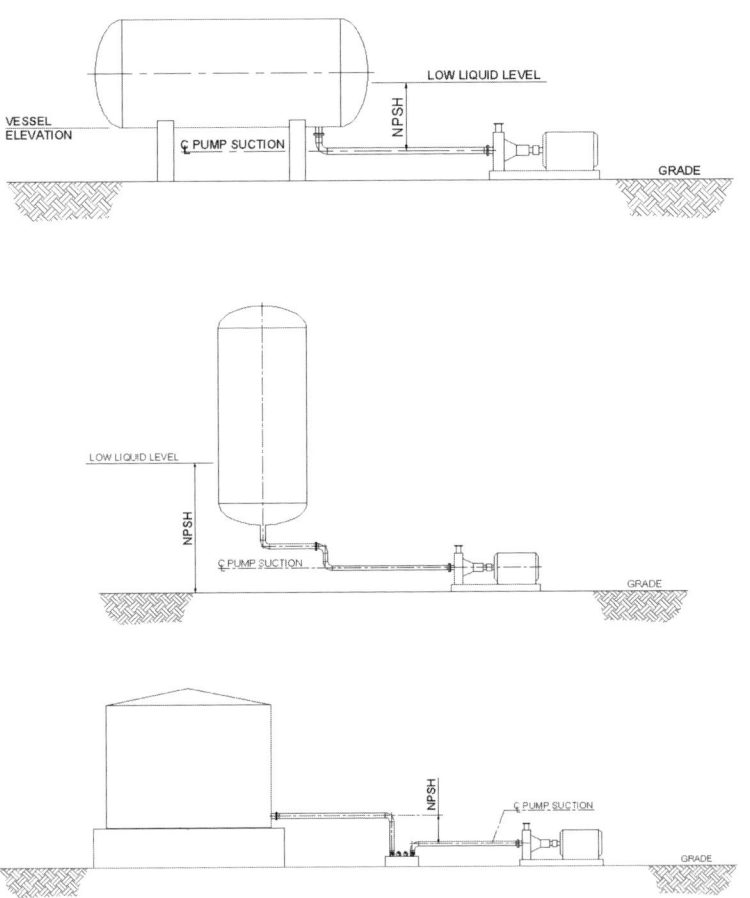

Figure 2–2   *Net positive suction head (courtesy of Red Bag/Bentley Systems, Inc.).*

Line Size

Suction piping usually is one or two line sizes larger than the pump suction nozzle size. Suction piping more than two sizes larger should be queried with the Process Department. For example, a 10" suction nozzle on a pump where the suction size of the pipe is 6" or 8" is probable, but the same suction nozzle where the suction size of the pipe is 3" or 4" is questionable, check with the Process Department.

Suction Nozzle Orientation

- Centrifugal pumps are supplied with suction nozzles on the end of the pump casing, axially in line with the impeller shaft; however, they also are on the top or side of the pump casing.
- Usually, pumps are specified with end or top suction for general services.
- Side-suction pumps, with side discharge, frequently are selected for large-volume water duty.
- Also side suction–side discharge pumps can be obtained in multistage form for higher pressure differentials. These pumps tend to become very long, so if plot space is tight, consideration should be given to purchasing the pump in a vertical form with a sump at grade. See Figure 2–3.
- Consider the use of flanged removable spool pieces, to allow pumps to be removed if required, without cutting the pipe and, therefore, avoiding additional fabrication.

Flexibility of Suction Lines

- Consistent with good piping practice, suction lines between the vessel and the pump should be as short as possible to avoid an unnecessary drop in pressure.
- Pump suction lines should be arranged so that unnecessary changes of direction are avoided and turbulence in the flow is minimized.
- Low points and pockets where vapor and gases can be trapped must be avoided.
- However, there must be sufficient flexibility in the piping system to absorb any pipe movement caused by temperature differentials or liquid surges and to maintain pump nozzle loads to within those permitted by pump vendor.

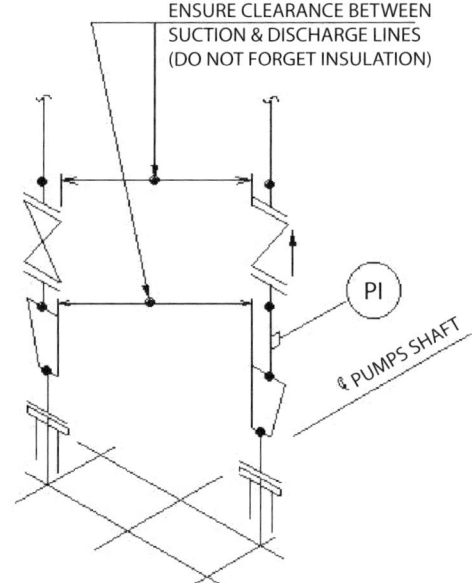

ENSURE CLEARANCE BETWEEN
SUCTION & DISCHARGE LINES
(DO NOT FORGET INSULATION)

PI

⅊ PUMPS SHAFT

TOP SUCTION - TOP DISCHARGE PUMPS

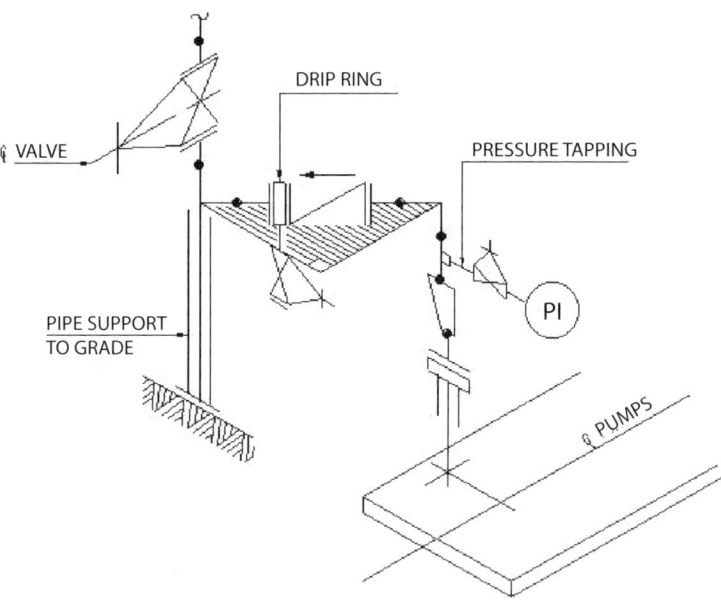

DRIP RING

⅊ VALVE

PRESSURE TAPPING

PIPE SUPPORT
TO GRADE

PI

⅊ PUMPS

Figure 2–3   *Suction and discharge piping clearances (courtesy of Red Bag/Bentley Systems, Inc.).*

Suction Line Fittings

- Due to the suction line being larger than the suction nozzles, reducers are required in the line. Reducers should be as close as possible to nozzle.

   Eccentric reducers are used with the flat on top for horizontal pumps. This prevents vapor being trapped and encouraging the phenomenon of cavitation. See Figure 2–4.

- For pumps with suction and discharge nozzles on the top of the casing, care must be taken to ensure that the flats on eccentric reducers are orientated back to back, so that suction and discharge lines do not foul each other. See Figure 2–4.

- Isolation valves must be provided on the suction line upstream of the strainer to allow its removal. It should be located within 3 m (10 ft) of the pump nozzle and should be accessible for hand operation.

- Generally, the size of the isolation valve is the same as the suction line.

- Casing and baseplate drains should be piped to the appropriate piping system.

Temporary Startup Strainers

- All pumps must have a temporary startup strainer in the suction line to prevent any pipe debris damaging the internals of the pump. The mesh size of the strainer must be specified or approved by the pump manufacturer, which is aware of the pump's characteristics.

- Strainers are located between the pump suction isolation valve and the pump. This allows the pump to be isolated and the strainer removed to be cleaned or replaced.

- Strainers are available in the following styles: flat, basket, conical, and bath or "tee" type.

- For the basket and conical types, a removable spool piece must be provided downstream of the suction block valve, which must not interfere with line supports. Both types have the advantage that the piping is left undisturbed and strainer element can be removed simply by unbolting the blind flange on the tee, thus leaving the piping and supports undisturbed. See Figure 2–5.

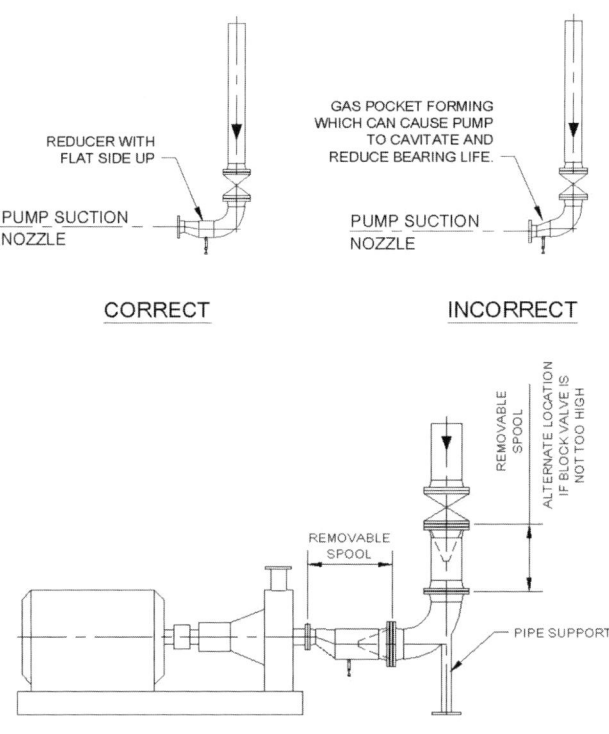

Figure 2–4 *Horizontal pump reducer positions (courtesy of Red Bag/Bentley Systems, Inc.).*

Discharge Piping

Line Size

Generally, discharge piping is one or two sizes larger than the pump discharge nozzle size. For example, a 10" suction nozzle on a pump where the suction size of the pipe is 12" or 14" is probable, but the same suction nozzle where the suction size of the pipe is 16" or 18" is questionable, check with the Process Department.

Discharge Line Piping Fittings

- Due to discharge lines being larger than the discharge nozzle, eccentric reducers are required in the line.
- Reducers should be as close as possible to the nozzle; with top suction–top discharge pumps, care must be taken to ensure

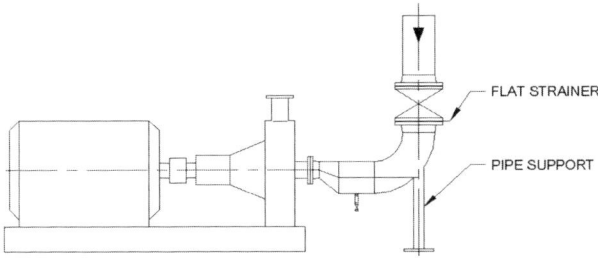

FLAT STRAINER

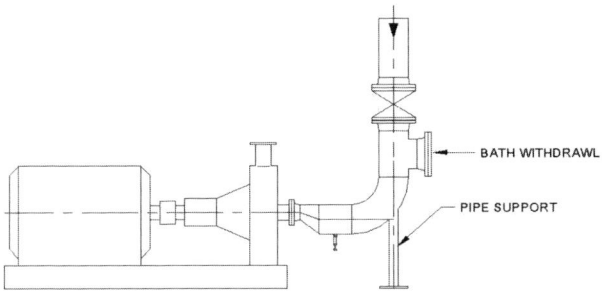

TEE TYPE STRAINER

Figure 2–4   *Continued.*

that the flats on eccentric reducers are orientated so that the lines do not foul each other.

- A pressure gauge should be located in the discharge line, upstream of the check and isolation valves.
- When a level switch for pump protection is installed in the discharge line, upstream of the block valves, ensure good access for maintenance of switch.
- To enable good access to valve handwheels and ease of supporting, the discharge line should be turned flat after the reducer, and the line angled away from the nozzle to enable the line to be supported from grade.
- Avoid supporting large lines from pipe-rack structures if possible, this enables minimum-size beam sections to be used and better access for pump removal and maintenance.

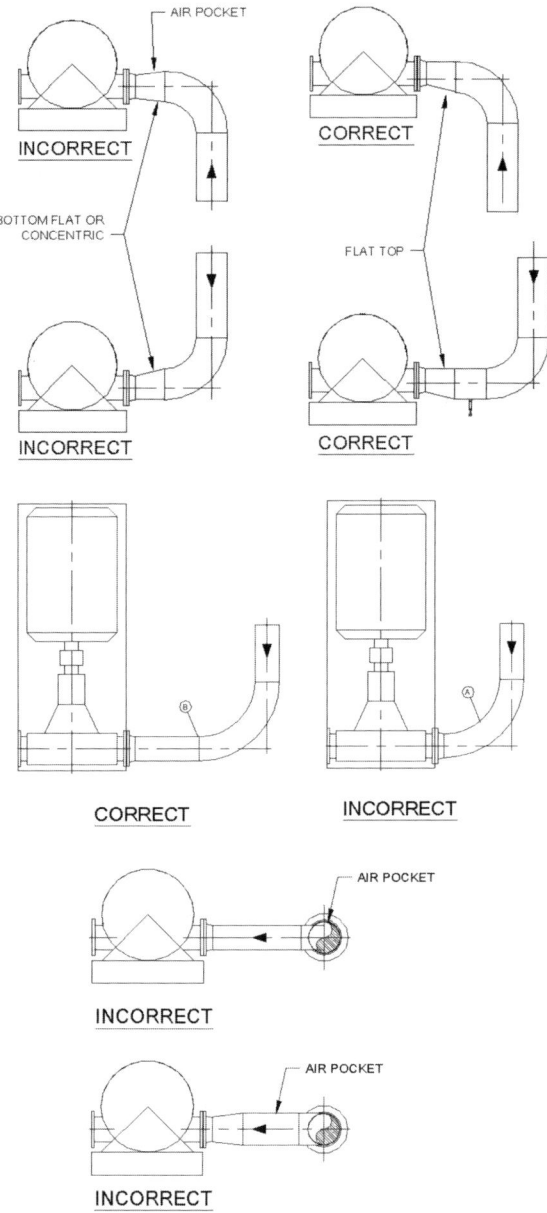

Figure 2–5   *Reducer and conical-basket strainer positions (courtesy of Red Bag/Bentley Systems, Inc.).*

Side-Suction and -Discharge Piping

- A horizontal centrifugal pump with side suction and discharge usually is installed for heavy duty service with large bore lines.

- Never connect an elbow flange fitting makeup to the nozzle of suction line coming down to the pump. Supply a straight piece of pipe two pipe diameters long between the nozzle and elbow.

- The two-diameter pipe length can be eliminated if the elbow is in the horizontal plane, eliminate the pipe length only if available space is tight.

2.5.2 Vertical Centrifugal Pump Piping

Vertical pumps, also called *can-type* or *barrel-type pumps*, are used when the available NPSH is very low or nonexistent.

Vertical In-Line Pumps

- The vertical in-line pump is mounted directly into the pipe line, as you would a valve.

- For smaller sizes, the piping system supports the pump and motor; therefore, it is essential that the line is supported local to the pump to prevent the line moving when the pump is removed.

- Ensure that there is good access to pump for maintenance and withdrawal with no overhead obstructions for lifting out pump.

- Larger-size in-line pumps have feet or lugs on the casing for supporting from grade or steelwork.

Vertical Can- or Barrel-Type Pumps

- Usually, this type of pump is installed in cooling tower water circulating service, retention ponds, and applications where the NPSH is low and suction is taken from a sump below grade.

- In most cases, there is no suction piping to be considered, but the discharge line must be routed to ensure good access for pump maintenance, with no overhead obstructions for pump removal by a crane. See Figure 2–6.

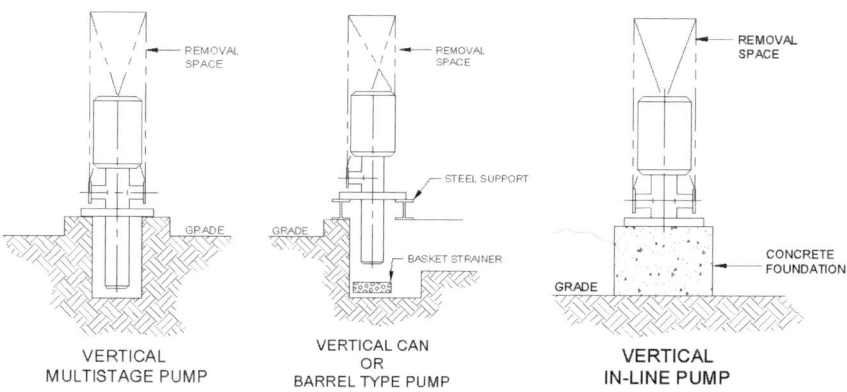

Figure 2–6   *Vertical pump piping arrangements (courtesy of Red Bag/Bentley Systems, Inc.).*

2.5.3   Reciprocating Pump Piping

- The reciprocating pump's discharge per stroke determines the quantity of liquid to be delivered in to the system.
- The suction and discharge lines of a reciprocating pump should have pulsation dampers installed where vibration might be caused in the pumps due to cyclic pulsation of the liquid.
- The location of dampers, if installed in the discharge piping of the pump, should be selected for the nearest location to discharge nozzle of pump to minimize the piping length.
- The piping layout should not cause any obstruction to the operation of the pump or any difficulty in the maintainability of the pump.
- A check valve usually is not required in the discharge piping to prevent back flow.

2.5.4   Rotary Pump Piping

Rotary pumps are used for viscous liquids free of solids or vapor. The liquid is pushed by means of gears or screws. These pumps used only in special cases. Each design case needs to be addressed separately, and the vendor of the pump should participate in the design process.

2.5.5   Steam Turbine Driver Piping

General

Rotor

Most small turbine casings are split along their horizontal axis and enough space above the turbine should be kept clear to allow for the top section of casing to be lifted clear of the rotor by a crane.

Steam Nozzle Orientation

- The steam inlet nozzle usually is on the right-hand side when viewed from the pump coupling end of the turbine, with the exhaust on the left-hand side.
- Turbines can be purchased with the inlet and the exhaust on the same side. This means that the piping designer can place the exhaust connection either on the same or opposite side from the inlet. Generally, opposite side location of nozzles results in less piping congestion.

Warm-up Bypass

- On automatic startup, a warm-up bypass must be provided around the control valve. This bypass is usually a 1" globe valve, is partially opened to allow steam to keep the turbine constantly warm, and is slowly turning to prevent the shock of hot steam entering a cold turbine, to eliminate damage to turbine blades.
- For manual startup, it is recommended that a warm-up bypass be installed, but the job flow sheets govern.
- When a warm-up bypass is installed, a steam trap on the casing keeps the system free of condensate.

Suction Piping

- Steam inlets are furnished with strainers as part of the turbine for protection against pipe debris; therefore, inlet piping must be designed with a removable section for strainer removal.
- Steam supply to turbines must avoid pockets in the line and be moisture free at all times; otherwise, the turbine will be damaged if condensate enters the turbine while it is running.
- To separate condensate from the steam, a trap must be installed upstream of the inlet block valve.
- The two basic turbine installations are manual startup and automatic startup. The manual startup has an isolation valve in the steam supply near the turbine inlet. Upstream of the

isolation valve, a bootleg must be installed with connections for blowdown and a steam trap to remove any condensate in the steam supply.

- For automatic startup, the isolation valve is replaced with a remote operated control valve; the bootleg and traps are still required upstream as for manual startup.
- Steam traps should be provided to keep the turbine casing free from condensate. These can be installed at the casing low point if a connection is provided or on the outlet piping if the casing drains into the outlet system.
- Note: There must be a trap before any vertical rise that could form a pocket where condensate can be collect and piped away to a collection system.

Discharge Piping
- Turbine exhausts are routed to either a closed exhaust steam system or the atmosphere.
- When an exhaust is to a closed system, there must be a block valve between the turbine and the main header. This block always is open during normal conditions and closed only for turbine maintenance or removal.
- Thought should be given to locating exhaust block valve on the pipe rack immediately before lines enter main header; this prevents accidental closure of this valve.
- If the exhaust line is routed to atmosphere, the steam trap on turbine casing is not installed but replaced by gate valve, partially open to allow condensate to drain off the casing.

2.6   Auxiliary Piping

- Most pumps require external services to be piped to them for bearing cooling, bearing lubrication, seal flushing, venting, and draining.
- These requirements are shown on utility flow sheets, and it is the piping designer's responsibility to ensure that the actual geographic location of pumps with harnesses are correctly shown on the flow sheets.
- Thought should be given to running subheaders to groups of pumps that have harness requirements. These subheaders must be sized and marked on flow sheet masters. Because branch lines to individual pumps are small in diameter, that is, 6 mm, it is advisable to take branch connections from the

top of the subheaders. This prevents pipe debris getting into the branch line, the pump bearings, and the like.

- Most vendors supply the auxiliary piping to the pump seals through a harness. Care should be taken to ensure harness piping does not interfere with good operation and maintenance space.

2.7 Piping Support and Stress Issues

- The piping around pumps should be designed to minimize the forces to the pumps caused by piping stresses.
- The stresses can be caused by temperature variation, piping weight with fluid, or vibrations in piping systems.
- The pumps are high-precision rotating machinery and can be damaged by misalignment in the casing or foundation of the pump.
- The allowable forces to the nozzles of the pumps should be determined according to the job or vendor specification.
- Each line connected to a pump should be considered for thermal stress calculation.
- The "line index/list" is used to make the first selection of lines to be calculated for thermal stress.
- The first selection usually is based on the operating temperature.
- Care should be exercised, since in some cases, the piping system does not present the same constant temperature, depending the mode of operation.
- Common expansion loops, but also flexible joints and ball joints, should be used to minimize the external forces due to thermal expansion and contraction.
- The various restrictions or free supports, such as anchor support, stopper, or tension rods, can be used to reduce the force on the nozzles of the pump.

CHAPTER 3

Compressors

3.1   Introduction

Compressors are the mechanical means to increase pressure and transport a vapor from one location to another, in the same way that a pump increases pressure to transport a liquid through a piping system. The following text is not intended to influence the selection of a compressor but to highlight certain issues that must be considered when laying out the suction and discharge pipework to the compressor.

3.2   Types of Compressors

Two basic types of compressors are used in process plants: reciprocating and centrifugal. Each type of compressor has the specific duty to take in the vapor at low pressure, compress it, and discharge the vapor at a higher pressure. The quantity of vapor to be moved and the discharge pressure usually are the deciding factors when selecting the type compressor to be used.

3.2.1   Reciprocating

Reciprocating compression is the force converted to pressure by the movement of the piston in a cylindrical housing. These machines generally are specified for transporting lower volumes of vapor than centrifugal compressors. If several stages of compression are employed, extremely high pressures can be achieved. Because of their reciprocating action, these machines cause piping systems that are not properly designed and supported to pulsate, vibrate, and generate fatigue that may result in fracture and system failure. Therefore, care should be taken with the materials of selection and method of jointing, and there must be sufficient flexibility in the piping system to keep the loads on the nozzles of the compressor to a level that is acceptable to the manufacturer's recommendations.

3.2.2 Centrifugal

Centrifugal compression is the force converted to pressure when a gas is ejected by an impeller at increasing velocity. Generally, centrifugal compressors are specified when large quantities of vapor have to be transported through the piping system. The suction-discharge pressure differential range is larger than that of reciprocating compressors. Centrifugal compressors are not subject to the same pulsation and vibration issues as reciprocating compressors and, therefore, do not produce the effects that may result in potential piping system failure.

3.3   Drivers

Drivers are required to power the compressor, and they fall into three categories: electric, steam, and gas. Electrical drivers range from small flameproof motors to large motors, 2000 hp or larger, that require an independent cooling system. Steam drivers comprise single- or multi-stage turbines, either fully condensing or noncondensing. Gas drivers cover gas turbines or gas internal combustion engines. The driver is selected based on several factors: safety, suitability, availability, and cost.

3.4   Applicable International Codes

Numerous international codes and standards that apply to the compressors and associated equipment are used in hydrocarbon processing plants, and the following are several of the most important, along with their scope and tables of contents.

The design and specifying of these items of equipment are the responsibility of the mechanical engineer; however, a piping engineer or designer benefits from being aware of these documents and reviewing the sections that relate directly to piping or a mechanical-piping interface. The standards discussed are

- API Standard 616.
- API Standard 617.
- API Standard 618.
- API Standard 619.

3.4.1 API Standard 616. Gas Turbines for the Petroleum, Chemical, and Gas Industry Services

Scope

This standard covers the minimum requirements for open, simple, and regenerative-cycle combustion gas turbine units for services of mechanical drive, generator drive, or process gas generation. All auxiliary equipment required for operating, starting, and controlling gas turbine units and turbine protection is either discussed directly in this standard or referred to in this standard through references to other publications. Specifically, gas turbine units that are capable of continuous service firing gas or liquid fuel or both are covered by this standard.

Table of Contents

1. Scope.
   1.1. Alternative Designs.
   1.2. Conflicts.
2. References.
   2.1. Referenced Standards.
   2.2. Compliance.
   2.3. Responsibilities.
   2.4. Unit Conversion.
3. Definitions.
4. Basic Design.
   4.1. General.
   4.2. Pressure Casings.
   4.3. Combustors and Fuel Nozzles.
   4.4. Casing Connections.
   4.5. Rotating Elements.
   4.6. Seals.
   4.7. Dynamics.
   4.8. Bearings and Bearing Housing.
   4.9. Lubrication.
   4.10. Materials.
   4.11. Nameplates and Rotational Arrows.
   4.12. Quality.
5. Accessories.

5.1. Starting and Helper Driver.

5.2. Gears, Couplings, and Guards.

5.3. Mounting Plates.

5.4. Controls and Instrumentation.

5.5. Piping and Appurtenances.

5.6. Inlet Coolers.

5.7. Insulation, Weatherproofing, Fire Protection, and Acoustical Treatment.

5.8. Fuel System.

5.9. Special Tools.

6. Inspection, Testing, and Preparation for Shipment.

6.1. General.

6.2. Inspection.

6.3. Testing.

6.4. Preparation for Shipment.

7. Vendor's Data.

7.1. General.

7.2. Proposals.

7.3. Contract Data.

Appendix A. Typical Data Sheets.

Appendix B. Gas Turbine Vendor Drawing and Data Requirements.

Appendix C. Procedure for Determination of Residual Unbalance.

Appendix D. Lateral and Torsional Analysis Logic Diagrams.

Appendix E. Gas Turbine Nomenclature.

3.4.2 API Standard 617. Axial and Centrifugal Compressors and Expander-Compressors for Petroleum, Chemical and Gas Industry Services

Scope

This standard covers the minimum requirements for axial compressors, single-shaft and integrally geared process centrifugal compressors, and expander-compressors for use in the petroleum, chemical, and gas industries services that handle air or gas.

This standard does not apply to fans (covered by API Standard 673) or blowers that develop less than 34 kPa (5 psi) pressure rise above atmospheric pressure. This standard also does not apply to packaged, integrally geared centrifugal plant and instrument air compressors, which are covered by API Standard 672. Hot gas expanders, over 300°C (570°F), are not covered in this standard.

Chapter 1 contains information pertinent to all equipment covered by this standard. It is to be used in conjunction with the following chapters as applicable to the specific equipment covered: Chapter 2, "Centrifugal and Axial Compressors"; Chapter 3, "Integrally Geared Compressors"; and Chapter 4, "Expander-Compressors."

Table of Contents

1. General Requirements.

2. Centrifugal and Axial Compressors.

3. Integrally Geared Compressors.

4. Expander-Compressors.

3.4.3 API Standard 618. Reciprocating Compressors for Petroleum, Chemical and Gas Industry Services

Scope

This standard covers the minimum requirements for reciprocating compressors and their drivers used in petroleum, chemical, and gas industry services for handling process air or gas with either lubricated or nonlubricated cylinders. Compressors covered by this standard are of moderate to low speed and in critical services. Also covered are the related lubricating systems, controls, instrumentation, intercoolers, aftercoolers, pulsation suppression devices, and other auxiliary equipment.

Compressors not covered are (a) integral gas-engine-driven compressors with single-acting trunk-type (automotive-type) pistons that also serve as crossheads and (b) either plant or instrument-air compressors that discharge at a gauge pressure of 9 bar (125 psi) or less. Also not covered are gas engine and steam engine drivers. The requirements for packaged reciprocating plant and instrument-air compressors are covered in API Standard 680.

Table of Contents

1. General.

    1.1. Scope.

    1.2. Alternative Designs.

    1.3. Conflicting Requirements.

    1.4. Definition of Terms.

5. Vendor's Data.

    5.1. General.

    5.2. Proposals.

    5.3. Contract Data.

Appendix A. Typical Data Sheets.

Appendix B. Required Capacity, Manufacturer's Rated Capacity, and No Negative Tolerance.

Appendix C. Piston Rod Runout in Horizontal Reciprocating Compressors.

Appendix D. Repairs to Gray or Nodular Iron Castings.

Appendix E. Control Logic Diagramming.

Appendix F. Vendor Drawing and Data Requirements.

Appendix G. Figures and Schematics.

Appendix H. Material Specifications for Major Component Parts.

Appendix I. Distance Piece Vent, Drain, and Inert Gas Buffer Systems for Minimizing Process Gas Leakage.

Appendix J. Reciprocating Compressor Nomenclature.

Appendix K. Inspector's Checklist.

Appendix L. Typical Mounting Plate Arrangement.

Appendix M. Pulsation Design Studies.

Appendix N. Guideline for Compressor Gas Piping Design and Preparation for an Acoustical Simulation Analysis.

Appendix O. Guidelines for Sizing Low Pass Acoustic.

Appendix P. Material Guidelines for Compressor Filters Components-Compliance with NACE MR0175.

Appendix Q. International Standards and Referenced Publications.

3.4.4 API Standard 619. Rotary-Type Positive-Displacement Compressors for Petroleum, Petrochemical and Natural Gas Industries

Scope

This standard covers the minimum requirements for dry and oil-flooded helical lobe rotary compressors used for vacuum or pressure or both in the petroleum, petrochemical, and natural gas industries. It is intended for compressors in special-purpose applications. It does not cover general-purpose air compressors, liquid ring compressors, or vane-type compressors.

Table of Contents

6.6. Intercoolers and Aftercoolers.

6.7. Inlet Air Filters.

6.8. Inlet Separators.

6.9. Pulsation Suppressors/Silencers for Dry Screw Compressors.

6.10. Special Tools.

7. Inspection, Testing, and Preparation for Shipment.

7.1. General.

7.2. Inspection.

7.3. Testing.

7.4. Preparation for Shipment.

8. Vendor's Data.

8.1. General.

8.2. Proposals.

8.3. Contract Data.

Annex A (Normative). Typical Data Sheets.

Annex B (Informative). Nomenclature for Equipment.

Annex C (Informative). Forces and Moments.

Annex D (Normative). Procedure for Determination of Residual Unbalance.

Annex E (Informative). Typical Schematics for Oil System for Flooded Screw Compressor.

Annex F (Informative). Materials and Their Specifications for Rotary Compressors.

Annex G (Informative). Mounting Plates.

Annex H (Informative). Inspector's Checklist.

Annex I (Informative). Typical Vendor Drawing and Data Requirements.

Bibliography.

3.5   Piping—Specific Guidelines to Layout

The following issues should be taken into consideration when piping up a compressor. Not all points are mandatory and many are based on common sense, safety, and best practices that will result in a safe, efficient, and economic piping system.

Special consideration is required in the design of a piping system at and near compressors to reduce fatigue failures and possible costly plant shutdown. The piping system should have the minimum over-hanging weight, and bracing should be provided as needed to reduce the vibration created by the compressor. The use of high-integrity butt welding fittings is recommended and should be considered instead of socket weld fittings.

3.5.1   Reciprocating

Introduction

Type of Machines

Reciprocating compressors can be obtained in a variety of patterns, from a simple single-cylinder to multicylinder, multistage machines. See Figures 3–1, 3–2, and 3–3 for the most widely used patterns.

Type of Cylinders

Figures 3–4 and 3–5 show details of the cylinder arrangements. Multicompression stages refer to the number of times the vapor is compressed by going through a series of compression cylinders to increase pressure.

When a gas or vapor is compressed, this raises the temperature of the product. In a reciprocating machine, compression is violent and the increase in the temperature significant. Inlet temperatures of 40°C may be raised to over 100°C by the act of compression. The cylinder becomes hot, and depending on the vapor being compressed, it will need some form of cooling. This usually is in the form of cooling water, but for low heat increases, a glycol-filled jacket may be specified.

Compressor Layout

Foundation

- The foundation for LP reciprocating compressors *must* be independent from all other foundations local to the machine. The foundation of the machine must be of a sufficient size so that it can support the compressor and all its auxiliary equipment.

- Cylinder supports are supplied by the vendor if they are required, and they must be attached to the compressor foundation concrete. Likewise, the snubber supports must be attached to the foundation of the compressor and springs can be used locally to support the snubbers.

Operation and Maintenance

An effective compressor layout (see Figure 3–6) results in cost savings on process and utility piping, good maintenance accessibility, and

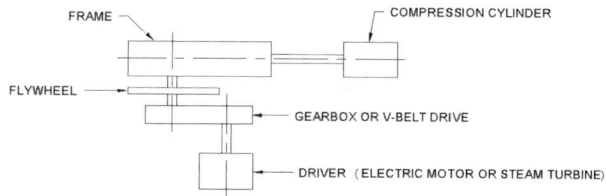

Figure 3–1 *A single cylinder machine (angle type). It will operate at low speed and may be single or double acting (courtesy of Red Bag/Bentley Systems, Inc.).*

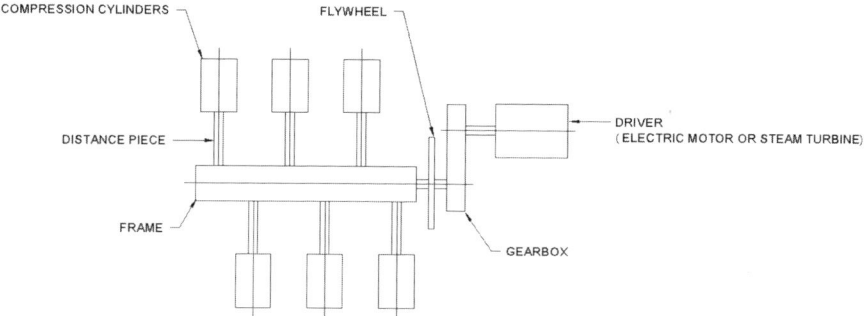

Figure 3–2 *A balanced horizontally opposed multicylinder machine. It will operate at low speed and may be single or double acting; it also can be multistage (courtesy of Red Bag/Bentley Systems, Inc.).*

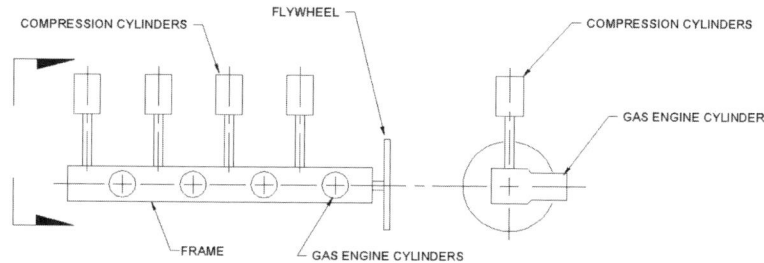

Figure 3–3 *A gas-fueled angle-type engine. All the compression cylinders are on one side of the frame and cylinder diameters and lengths vary according to the composition, pressure, and volume of gas to be compressed. Dimensions from frame center line to cylinder nozzles vary with compression forces. (Courtesy of Red Bag/Bentley Systems, Inc.) Note: Gas engine may take a V form.*

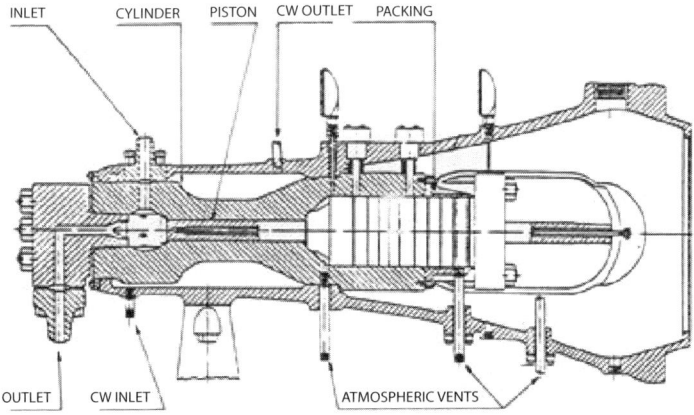

Figure 3–4   *Single-acting cylinder having one suction, compression, and discharge area per cylinder (courtesy of Red Bag/Bentley Systems, Inc.).*

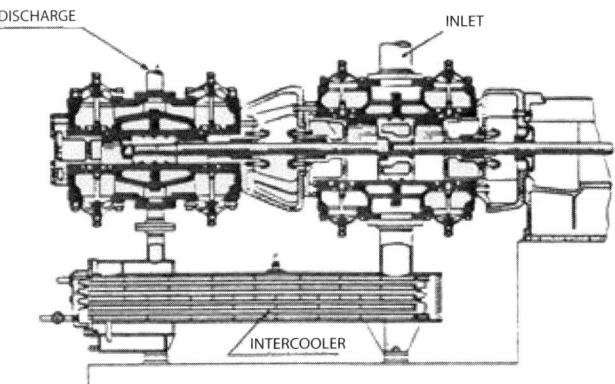

Figure 3–5   *Double-acting cylinder having two suction, compression, and discharge areas per cylinder (courtesy of Red Bag/Bentley Systems, Inc.).*

reduced pulsation and vibration in suction and discharge piping. Poor piping layout may fail and be responsible for a costly shut down.

- For angle-type compressors, locate the crankshaft parallel to the suction and discharge headers.

- For balanced, horizontally opposed compressors, the crankshaft should run at right angles to the suction and discharge headers.

- Compressor houses containing more than one machine, particularly if they are long, probably are equipped with a

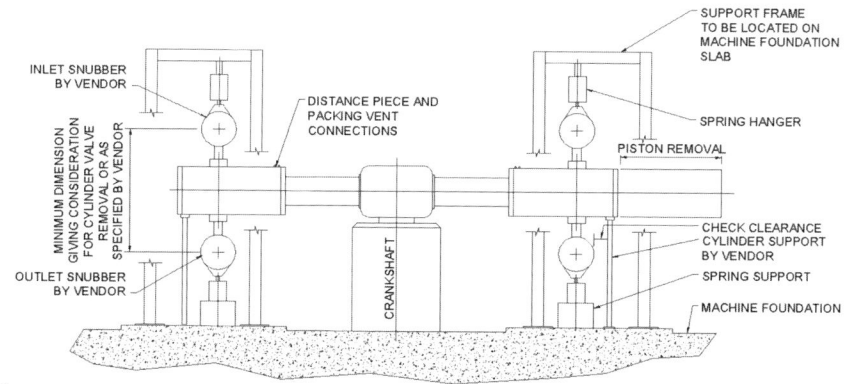

Figure 3–6 *Typical layout of compressor house and suction knockout drum (courtesy of Red Bag/Bentley Systems, Inc.).*

traveling gantry crane, which is operated manually or electrically.

- This feature can influence the overall dimensions of the house, as in addition to the necessary building and maintenance clearances, the vertical reactions of the loaded crane increase the foundation size. Since these must not be connected to the machine foundations, the building size is affected.

- It is usual for compressor vendors to indicate the overall foundation dimensions on their layout drawings. These dimensions should be requested as early as possible and forwarded to the civil engineering group.

- The compressor building must be sized very early in the layout stage, even if only preliminary dimensions are available.

- When the dimensions of the compressor have been determined, add to these dimensions adequate clearance for maintenance plus possible control valve stations, lube oil equipment, local control panel, and the like.

- Allow at least 2 m all around the original dimensions. In practice, this 2 m allowance provides a walkway of only 1200–1500 mm, due to other items occupying floor space. With two or more machines, allow at least 2 m between compression cylinders to allow for adequate piston removal.

- All dimensions must be confirmed from certified vendor drawings.

- Allow a maintenance area at one end of the building. A 6 m bay should be sufficient.
- Pits, trenches, and similar gas traps should be avoided in gas compressor houses.
- Large reciprocating gas compressors usually are elevated above grade, with the mezzanine floor level with the top of the foundation for operation and maintenance.
- The height of the mezzanine floor above grade is kept to a minimum consistent with the adequacy of space for piping and access, especially to valves and drains.

Suction and Discharge Piping

Special consideration is required in the design of piping system at and near compressors to reduce fatigue failures and possible costly plant shutdown. The piping system should have the minimum over-hanging weight, and bracing should be provided as needed to reduce vibration created by the compressor. The use of high-integrity butt welding fittings is recommended and should be considered instead of socket weld fittings.

- Compressor suction piping should be suitably clean to avoid ingress of foreign material.
- The piping layout must follow the sequence on the process flow diagram (PFD) or process and instrument diagrams (P&IDs), as issued for the project. If they conflict with any of the following notes, the PFDs or P&IDs always take precedence.
- It is usual for the suction piping to be routed to the top of the cylinder and discharge piping from the bottom.
- Liquids must be prevented from entering the compressors. As liquids do not compress, extensive precautions must be taken to ensure that absolutely no liquid enters the compressor cylinder; a small quantity could do extensive damage and cause an unnecessary shutdown.
- If there is any doubt that the vapor is near its dew point, the suction line must be steam or electrically traced between the suction drum and the compressor inlet or local to the compressor inlet. The Process Department advises the extent of the tracing, and it will be shown on the flow diagram.
- Suction and discharge headers should be located at grade level on sleepers up to the first piece of connecting equipment, such as a suction knockout drum or aftercooler.

- Branch connections to the compressor from the suction header are taken from the top of the header.
- Suction and discharge piping should be kept as straight as possible between the compressors and headers.
- The use of short-radius bends or tees and similar installations giving opposed flow are permitted.
- The suction piping should be no less than the compressor nozzle size.
- Piping local to cylinders should have sufficient distance to permit proper maintenance on the cylinder valves.
- When compressors are elevated with a mezzanine floor, piping and valves normally are run under the floor.
- When more than one compressor is employed on the same service, all piping to and from the compressors should be valved so that an individual compressor may be shutdown and taken out of service.
- Spectacle blinds are installed at the compressor side of the isolating valves.
- Startup bypasses should be installed between the suction and discharge pipes of compressors and located between the compressor and the line block valve.
- If a relief valve is not supplied by the compressor manufacturer, then one should be installed between the compressor discharge and block valve. This relief valve should discharge into the suction line downstream of the block valve. The relief valve is provided with a bypass for hand venting.
- Distance piece and packing vent piping are manifolded into systems as indicated on the flow diagrams. These systems should either vent to the atmosphere outside the compressor house or connect to a collection system.
- Utility piping comprises the cooling water supply and return to lube oil cooler, also cylinder jackets.
- The minimum pipe size usually is ¾" nominal pipe size.
- Sufficient vents and drains are provided so that water lines and jackets may be completely drained at shutdown.
- A steam or electrical supply may be required if lube oil heaters are provided for either the compressor or gear box oil. This system is used prior to startup.
- Check for lines that have to be chemically cleaned and ensure drawings indicate this requirement.

3.5.2   Centrifugal

Introduction

Types of Machines

Centrifugal compressors can be obtained in a variety of patterns. In centrifugal radial compressors (see Figure 3–7), the compression process is effected by rotating impellers of radial flow design (see Figure 3–8) in the fixed guide elements. In centrifugal axial compressors (see Figure 3–9), the force is converted into pressure by rotating vanes between fixed guide vanes; the flow is axial.

Size and Position of Nozzles

Centrifugal compressor manufacturers have basic case designs, and they change the rotor blade design to meet specific volume and pressure requirements. For this reason, suction nozzles sometimes are much larger or smaller than the line size for hydrocarbon process applications. For example, a 30" suction nozzle may have a 20" or 24" suction line. It is necessary to increase the suction line diameter locally at the compressor nozzle. Do not use a reducing a flange, as this will introduce full velocity to the rotor blades at a turbulent condition. Use a 30" flange and a concentric reducer as a minimum. It is better if a pipe length of three diameters of 30" pipe can be accommodated as well.

Suction and discharge nozzles are on either the underside or the top of the compressor. In multistage compressors, two or more inlet nozzles may be provided; the suction lines are connected to suction drums controlled to maintain the various inlet pressures.

Foundation

The foundation of each machine is combined with its direct coupled drives but must be independent from all other local foundations, including the lube console, see Figure 3–9.

Suction and Discharge Piping

- Centrifugal compressors usually are large capacity machines.
- They are driven by electric motors, steam, or gas turbines; the power may be via a gearbox.
- It is usual to mount such machines on a tabletop approximately 4 m high with elevated access all around.
- The lube and seal oil consoles for both the compressor and turbine, if required, usually are located at grade.

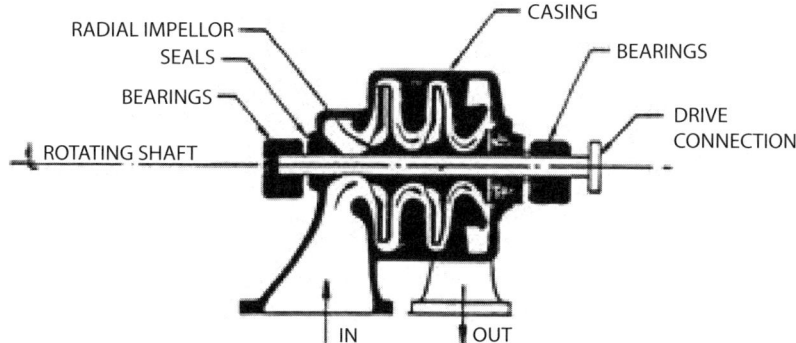

Figure 3–7   *Centrifugal radial compressor (courtesy of Red Bag/Bentley Systems, Inc.).*

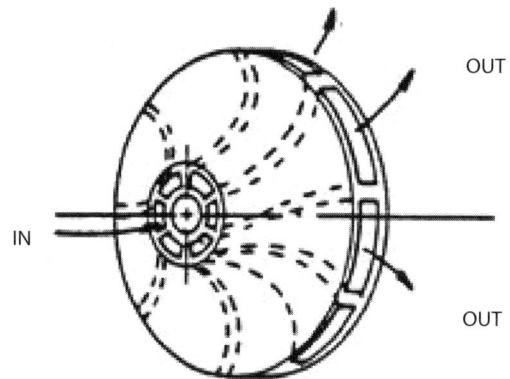

Figure 3–8   *Radial impeller (courtesy of Red Bag/Bentley Systems, Inc.).*

- A typical compressor house layout is shown in Figures 3–10 and 3–11. Here, an electrical motor and a condensing-type turbine have been used. Note the withdrawal and maintenance areas, also the acoustic hoods.

- The suction and discharge connections of the compressor most likely are on the underside; these lines can be anchored at grade. Should these connections be on the top of a horizontally split case compressor, see Figure 3–12 for details on removable spools.

- Determine the type of traveling gantry crane, and ensure that piping and so forth are clear of it. Note the lube oil header tanks, these must be elevated above the machines, if the

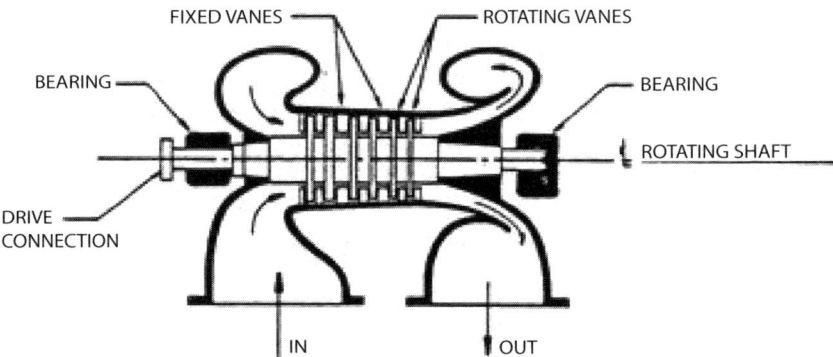

Figure 3–9   *Centrifugal axial compressor (courtesy of Red Bag/Bentley Systems, Inc.).*

vendor has not stated a minimum elevation use 10 m above the center line of the machines.

- The purpose of the tanks is for emergency lubrication, and they are tripped if the normal lubrication supply system should fail. They supply oil to the bearings until the machine comes to a standstill.

- The lube and seal oil consoles comprise the following items: oil storage tank, filters, pumps, oil cooler, sometimes an oil heater for startup, and control instruments.

- Interconnecting piping must be in accordance with the PFD and P&IDs.

- All return lines must be free draining from the machines to the console.

- Suction and discharge piping must be supported so that the nozzles are not overloaded, use reducers not reducing flanges local to suction and discharge nozzles.

- Compressor suction piping should be suitably clean to avoid ingress of solids.

- A temporary strainer should be installed on the suction piping.

- Make provision for removal of strainers in the suction piping with a spool piece.

- Silencers may be required in both the suction and discharge piping.

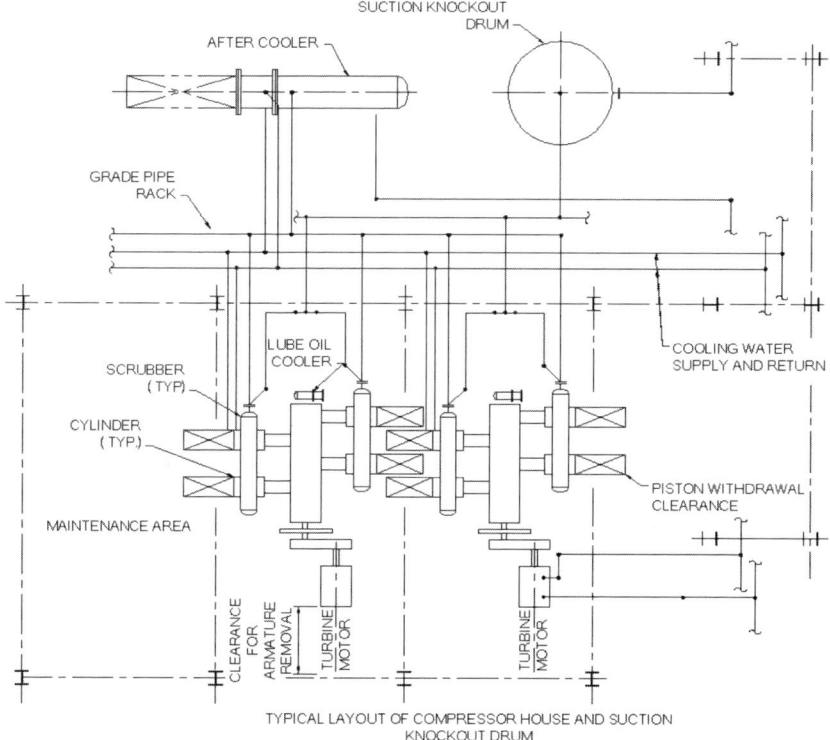

Figure 3–10 *Typical layout for compressors, one turbine driven and one electric motor driven (courtesy of Red Bag/Bentley Systems, Inc.).*

- Acoustic hoods may be required for both the compressor and turbine; ensure that the tabletop is large enough to accommodate these hoods.

- The hoods may be of sectional construction. The traveling gantry crane is used to dismantle them; this must be taken into consideration when determining the elevation of the crane hook.

- The maintenance area must be large enough to accommodate the acoustic hood, turbine and compressor half casing rotors, and so on.

3.5.3   Drives and Auxiliary Piping

Electric Motors

- Flameproof motors are employed for small- to medium-horsepower machines.

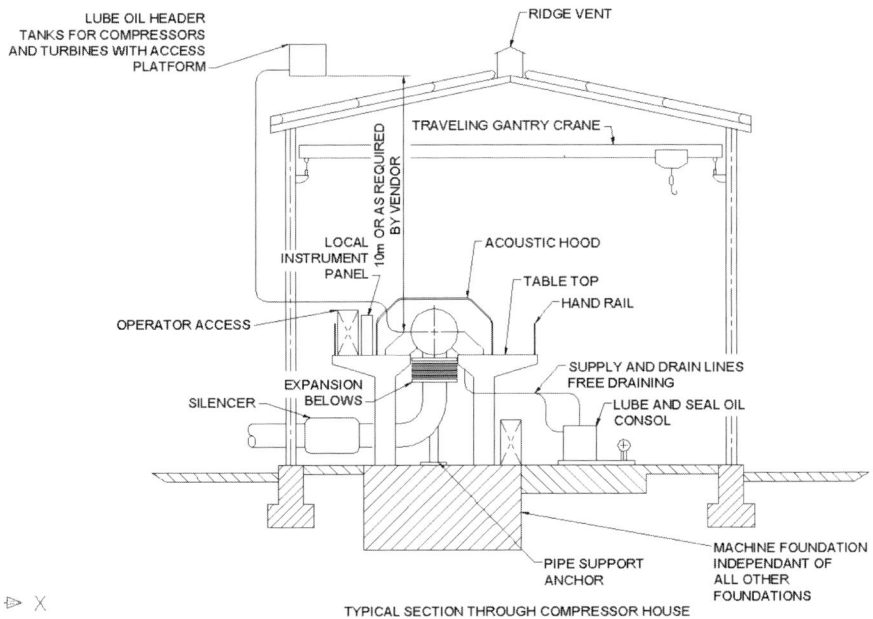

Figure 3–11   *Typical section through a compressor house (courtesy of Red Bag/Bentley Systems, Inc.).*

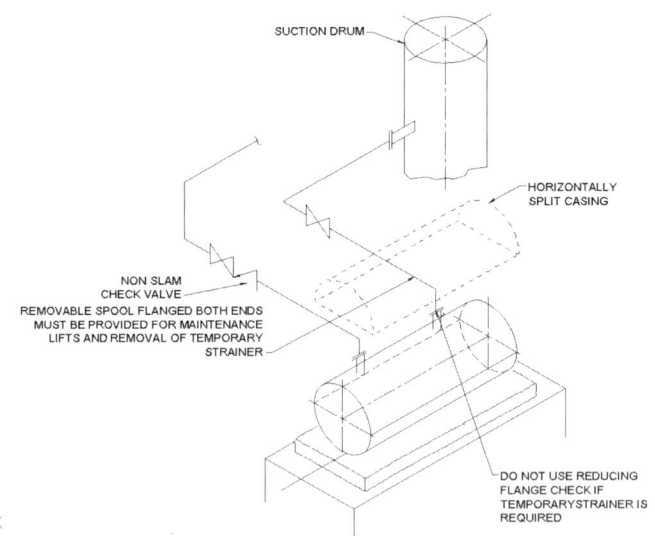

Figure 3–12   *The nozzle orientation for a horizontally split compressor casing (courtesy of Red Bag/Bentley Systems, Inc.).*

- Ensure that the cables can be routed to the terminations, also that there is space behind the motor to remove the rotor.
- Large, high-pressure machines with their own cooling systems fall into two categories: closed-air circulation, water-cooled (CACW) machines and closed-air circulation, air-cooled (CACA) machines.
- These types of machines may require a larger area; therefore, they affect the size of the compressor house.
- CACW machines (see Figure 3–13) may be mounted on a tabletop with the cooler located underneath; the cooler is located in a sealed room.
- The cooling air circulating around the motor is itself cooled by a water-cooled heat exchanger. Provision must be made for removal and service of the exchanger.
- It is possible to obtain motors with the cooler mounted above or to one side of the motor.
- For a CACA machine (see Figure 3–14), consideration must be given regarding the safe location of the air intake, which is outside the compressor house.
- If a filter is required in the intake system, provide access for replacement or cleaning.
- The lube and seal oil consoles comprise the following items: oil storage tank, filters, pumps, oil cooler, sometimes an oil heater for startup, and control instruments.
- Interconnecting piping must be in accordance with the PFD/P&ID.
- All return lines must be free draining from the machines to the console.

Steam Turbines

Two types of steam turbines must be considered, condensing and noncondensing.

- Steam driver piping, including drains, should be designed to avoid pockets (low points) to minimize the accumulation of condensation.
- Provision should be considered to bleed warming steam into turbines and other steam drivers.
- The noncondensing type of steam turbine uses high-pressure steam and exhausts a lower-pressure steam to a header.

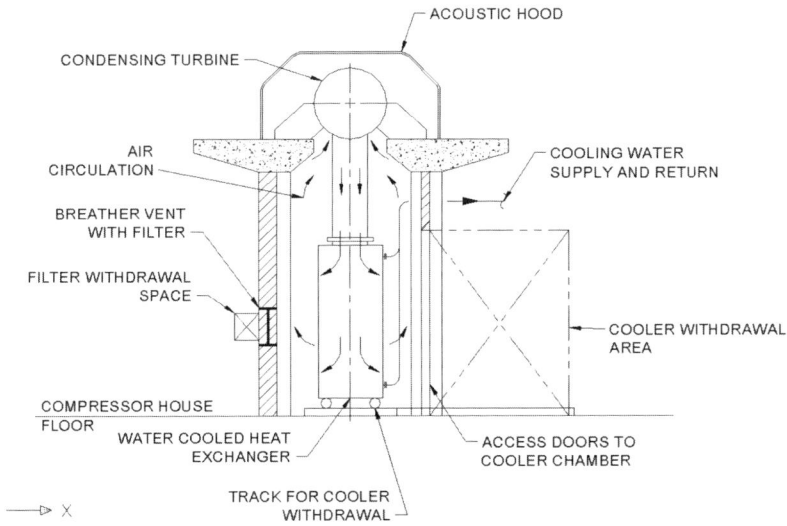

Figure 3–13   *Typical section through a closed-air-circulation, water-cooled machine (courtesy of Red Bag/Bentley Systems, Inc.).*

- The condensing turbine exhausts to a surface condenser, which usually is a large exchanger with a hot well attached but may take the form of an air fan, to recover condensate.
- Surface condensers often are at grade, mounted directly below the compressors turbine. This arrangement employs a turbine with an outlet nozzle directly connected via an expansion joint to the surface condenser (See Figure 3–15).
- The surface condenser may be mounted at grade alongside a grade-mounted turbine. With this arrangement very little NPSH is available.
- If an air fan is used as a surface condenser, it usually is located above the turbine, either on the compressor house roof or over a pipe rack.
- If the condenser is of the shell and tube type, it most likely has a fixed tube plate design and requires access for rodding the tubes.
- The cooling water lines associated with the condenser are large bore; and some consideration must be given to the piping arrangement and placing of valves to give good operation and utilization of plot space.

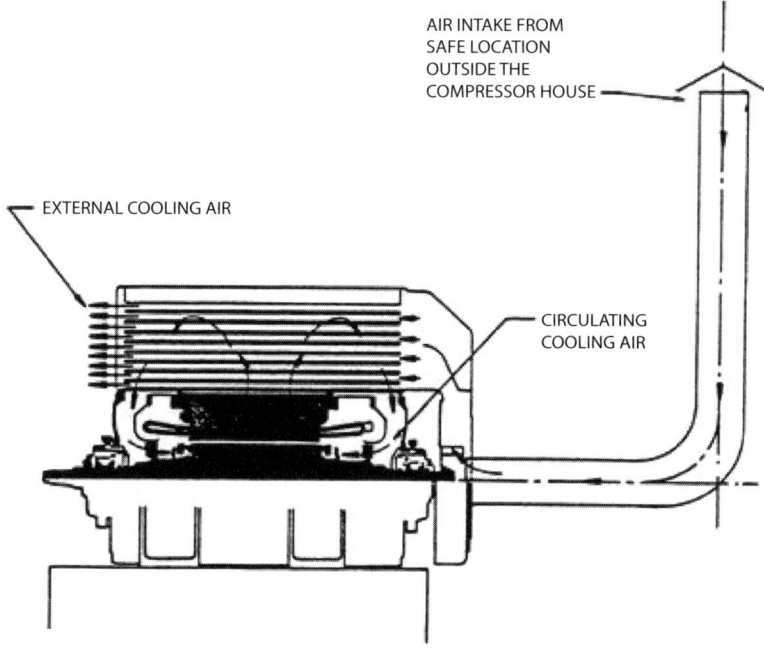

AIR INTAKE FROM
SAFE LOCATION
OUTSIDE THE
COMPRESSOR HOUSE

EXTERNAL COOLING AIR

CIRCULATING
COOLING AIR

Figure 3–14 *Typical section through a closed-air circulation, air-cooled machine. Note: On the CACA enclosure, a top-mounted air-to-air heat exchanger is used. The external air is circulated by means of a shaft-mounted fan in the case of cage machines and separate motor-fan units mounted in the ducting for wound rotor motors. (Courtesy of Red Bag/Bentley Systems, Inc.)*

- The steam supply to the turbine is taken from the top of the steam header, a bellow may be required local to the turbine and a temporary strainer for startup.

- The turbine requires a similar lube oil console to that provided for the compressor. Do not pocket the return drains. An elevated lube oil header tank also is required.

- Noncondensing turbine assemblies comprise a turbine, lube oil console, and header tank.

- The low-pressure steam discharge line has a large bore; a bellow most likely is required in the line, which must join the top of the header.

- If the line has a low point, a steam trap and drip pocket must be provided.

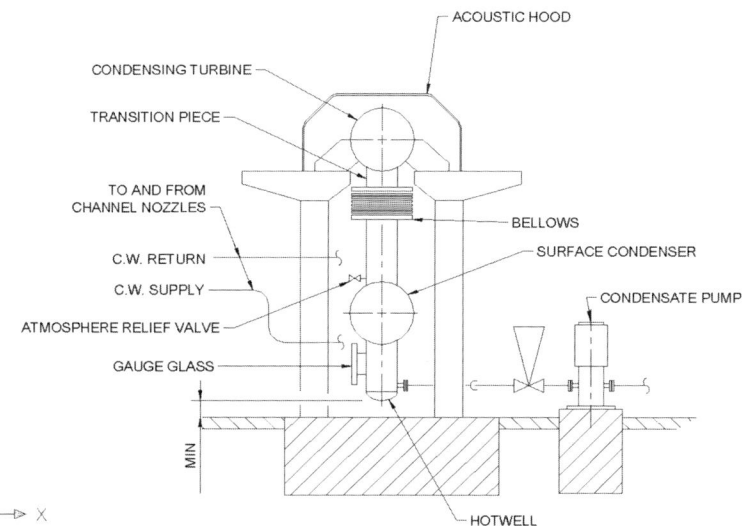

Figure 3–15   *Typical section through a condensing turbine set (courtesy of Red Bag/Bentley Systems, Inc.).*

- For maintenance access, provision must be made to dismantle the acoustic hood and remove half the turbine casing and the rotor.

Gas Turbines

- When using a gas turbine to drive a compressor, an arrangement similar to a steam turbine should be considered.
- The lube oil console and header tanks are required for the auxiliary piping system.
- In addition, the exhaust system must be considered; this comprises ducting to some heat recovery system, either a steam raising plant or process heaters.
- Combustion air to the turbine burner must be taken from a safe location outside the compressor house. An inlet silencer and filter most likely are required.
- Provision for operation and maintenance to all machinery must be provided.

Gas Engines

- Gas engines are used to drive reciprocating compressors, either directly or through a gearbox.

- The machine may have both compression and a drive cylinder attached to a common crankshaft. These types of engines may develop 2000 hp or more.
- Ensure that adequate space is allowed for removal of cylinder heads and pistons.
- The lube oil system may be integral to the engine or in the form of a console. Should the latter be used, ensure that the engine is at a suitable elevation to allow for free-draining oil return lines (see Figure 3–16).
- The engine and compressor are mounted on a common foundation, independent of all other foundations. Due to the vibration produced by these machines, a large mass concrete foundation is employed.
- The general layout of the compressor house should enable the use of a traveling gantry crane for all maintenance; therefore, when routing piping, this must be considered.
- Combustion air must be taken from a safe location outside the compressor house. If an air filter is required, arrange for maintenance access.
- Likewise, the exhaust must be discharged outside the building.
- This system should be fitted with a silencer and flame trap. Utility systems comprise a startup air system and fuel gas. The engine most likely has a closed-circuit jacket water cooling system, comprising a shell and tube exchanger or an air fan. If the former, a cooling water supply is required and the usual clearance for tube pulling and so forth.

3.5.4 Piping Support and Stress Issues

- Thermal expansion and dynamic forces are to be considered when designing the piping layout and pipe supports.
- All types of compressors compress vapor or gas, which increases the temperature of the gas. The temperature variation causes expansion of the piping.
- Vibration forces are the result of the inertial forces due to the weight of reciprocating elements and the balance weight of the rotating elements. Such forces act in the same plane and parallel to the axis of the cylinder bores.

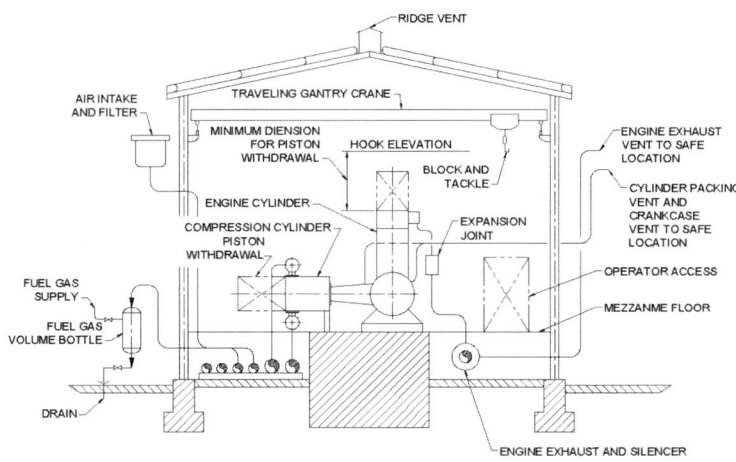

Figure 3–16 *Typical layout for free-draining utility lines (courtesy of Red Bag/Bentley Systems, Inc.).*

- The piping system connection usually starts with damper components that absorb the vibrational forces. However, vibration needs to be considered for the piping system in the direct vicinity of the reciprocating compressor and extra fixation of the piping is necessary.

CHAPTER 4

Exchangers

4.1   Introduction

A heat exchanger is a piece of equipment used to transfer heat from one media to another. In the petrochemical industry, heat exchangers generally can be classified under the following headings:

- Exchanger. Heats one stream and cools the other. There is no heat loss or physical change in either flowing media.

- Cooler. Cools liquid or gases without condensation; the term also applies to intercoolers and aftercoolers.

- Condenser. Condenses vapor or vapor mixtures. Can be water cooled or by sufficiently cold process stream, which requires heating.

- Chiller. Uses refrigerant to cool a process stream below the freezing point or below the prevailing cooling water temperature.

- Heater (nonfired). Heats process stream, generally up to its boiling point, without appreciable vaporization. The heating medium usually is steam or hot oil; the term also applies to preheaters.

- Reboiler. Reboils the bottom stream of the tower for the fractionation process. The eating medium can be steam or hot process stream. When large quantities of vapor have to be produced, a kettle-type reboiler is used.

- Waste heat boiler. Uses waste heat, such as internal combustion exhaust from gas turbines or similar drivers, to generate steam.

- Steam generator. Uses the heat of the process liquid or gas to produce steam.

- Vaporizer. Vaporizes part of a process liquid stream as does an evaporator.

4.2   Types of Exchangers

Exchangers can be divided into three groups: shell and tube, fin tube, and air fins.

4.2.1   Shell and Tube

Shell and tube exchangers can be vertical or horizontal, with the horizontal ones single or stacked in multiunits. As the name suggests, they consist of a cylindrical shell around a nest of tubes. The shell and tube exchangers can be further subdivided in three categories: floating head, U-tube, and fixed head.

Floating Head Shell and Tube Exchangers

Floating head exchangers are used when the media being handled causes fairly rapid fouling and the temperature creates expansion problems. Tubes can expand freely, the channel head and shell cover arrangement are convenient for inspection, and the tube bundle can be removed easily for cleaning.

U-Tube Shell and Tube Exchangers

U-tube exchangers are used when fouling of the tubes on the inside is unlikely. The tubes are free to expand and the bundle can be removed from the shell for cleaning the shell side of the tubes.

Fixed Head Shell and Tube Exchangers

Fixed head exchangers have no provision for the tube expansion and, unless a shell expansion joint is provided, can be used for only relatively low-temperature service. The end covers are removable, so that the inside of the tubes can be cleaned by rodding or using similar tools. This type of cleaning usually is carried out in situ, so some space should be allowed in the piping layout for this.

4.2.2   Fin Tube

Fin tube exchangers consist of a finned tube through which passes one media jacketed by another tube through which passes the other media. They can be used as single or multiunits.

4.2.3   Air Fin

Air fin exchangers come in two shapes, box type units and A-frame units. Both consist of banks of finned tubes through which passes the media to be cooled. Large fans blow air from the atmosphere through the banks, thus cooling the flowing media. Other types, such as plate exchangers and carbon block exchangers, are used infrequently.

Box Type Air Fin

The box type comes in two forms, forced draught and induced draught. Forced draft air fins are the more commonly used type, possibly because maintenance of the fan is easy from an underslung platform.

A-Frame Air Fin

A-frame air fins are less common than the box type. They offer the advantage of requiring less plot area than box-type air fins of the same capacity. They do present a few problems, however. Due to their physical shape, that is, a triangular section with the apex uppermost, the inlet header is located at the apex, with the collecting headers at both bottom corners. This means that cooled product lines come off both sides of the rack, which can present piping problems. Also, with 60° sides containing the product, it is possible to get uneven cooling, due to the sun being on one side or the prevailing wind tending to blow into the tube bank against the fan.

4.3   Applicable International Codes

Numerous international codes and standards apply to the heat exchangers used in hydrocarbon processing plants and several of the most important ones are presented here, along with their scope and table of contents.

The design and specifying of these items of equipment are the responsibility of the mechanical engineer; however, a piping engineer or designer will benefit from being award of these documents and reviewing the appropriate sections that relate directly to piping or a mechanical-piping interface.

- API Standard 660.
- ANSI/API Standard 661.
- ANSI/API Standard 662.

4.3.1 API Standard 660. Shell-and-Tube Heat Exchangers for General Refinery Services and ISO 16812:2002(E). Petroleum and Natural Gas Industries—Shell and Tube Heat Exchangers

Scope

This international standard specifies requirements and gives recommendations for the mechanical design, material selection, fabrication, inspection, testing, and preparation for shipment of shell-and-tube heat exchangers for the petroleum and natural gas industries. The standard is applicable to the following types of shell-and-tube heat exchangers: heaters, condensers, coolers, and reboilers. This standard is not applicable to vacuum-operated steam surface condensers and feed-water heaters.

Table of Contents

Annex C (Informative). Responsibility specification Sheet.

Annex D (Informative). Recommended Practices.

Bibliography.

4.3.2   ANSI/API Standard 661. Air-Cooled Heat Exchangers for General Refinery Service and ISO 13706-1:2005 (Identical). Petroleum, Petrochemical and Natural Gas Industries—Air-Cooled Heat Exchangers

Scope

This international standard gives requirements and recommendations for the design, materials, fabrication, inspection, testing, and preparation for shipment of air-cooled heat exchangers for use in the petroleum and natural gas industries. The standard is applicable to air-cooled heat exchangers with horizontal bundles, but the basic concepts also can be applied to other configurations.

Table of Contents

Foreword.

Introduction.

1. Scope.

2. Normative References.

3. Terms and Definitions.

4. General.

5. Proposals.

6. Documentation.

    6.1. Approval Information.

    6.2. Final Records.

7. Design.

    7.1. Tube Bundle Design.

    7.2. Air-Side Design.

    7.3. Structural Design.

8. Materials.

    8.1. General.

    8.2. Headers.

    8.3. Louvres.

    8.4. Other Components.

9. Fabrication of Tube Bundle.

    9.1. Welding.

    9.2. Post-Weld Heat Treatment.

    9.3. Tube-to-Tubesheet Joints.

    9.4. Gasket Contact Surfaces.

    9.5. Thread Lubrication.

    9.6. Alignment and Tolerances.

    9.7. Assembly.

10. Inspection, Examination and Testing.

    10.1. Quality Control.

    10.2. Pressure Test.

    10.3. Shop Run-In.

    10.4. Equipment Performance Testing.

    10.5. Nameplates.

11. Preparation for Shipment.

    11.1. General.

    11.2. Surfaces and Finishes.

    11.3. Identification and Notification.

12. Supplemental Requirements.

    12.1. General.

    12.2. Design.

    12.3. Examination.

    12.4. Testing.

Annex A (Informative). Recommended Practices.

Annex B (Informative). Checklist, Data Sheets and Electronic Data Exchange.

Annex C (Informative). Winterization of Air-Cooled Heat Exchangers.

Bibliography.

4.3.3 ANSI/API Standard 662. Plate Heat Exchangers for General Refinery Services, Part 1, Plate and Frame Heat Exchangers and ISO 15547-1:2005 (Identical). Petroleum, Petrochemical and Natural Gas Industries—Plate Type Heat Exchangers, Part 1, Plate and Frame Heat Exchangers

Scope

This part of API 662/ISO 15547 gives requirements and recommendations for the mechanical design, materials selection, fabrication, inspection, testing, and preparation for shipment of plate-and-frame heat exchangers for use in the petroleum, petrochemical, and natural gas industries. It is applicable to gasketed, semi-welded, and welded plate-and-frame heat exchangers.

Table of Contents

4.3.4 ANSI/API Standard 662. Plate Heat Exchangers for General Refinery Services, Part 2, Brazed Aluminum Plate-Fin Heat Exchangers and ISO 15547-2:2005 (Identical). Petroleum, Petrochemical and Natural Gas Industries—Plate Type Heat Exchangers, Part 2, Plate-and-Frame Heat Exchangers

Scope

This part of ISO 15547 gives requirements and recommendations for the mechanical design, materials selection, fabrication, inspection, testing, and preparation for shipment of brazed aluminum plate-fin heat exchangers for use in the petroleum, petrochemical, and natural gas industries.

Table of Contents

API Foreword.

Foreword.

Introduction.

1. Scope.

2. Terms and Definitions.

3. General.

4. Proposal Information Required.

5. Drawings and Other Data Requirements.

6. Design.

7. Materials.

8. Fabrication.

9. Inspection and Testing.

10. Preparation for Shipment.

Annex A (Informative). Recommended Practice.

Annex B (Informative). Plate-Fin Heat Exchanger Checklist.

Annex C (Informative). Plate-Fin Heat Exchanger Data Sheets.

Bibliography.

4.4   Piping—Specific Guidelines to Layout

4.4.1   Shell and Tube Exchangers

Exchanger Layout

- For good maintenance and safe working conditions, it is necessary to space exchangers such that the surrounding area is adequate and clear for accessibility (see Figure 4–1).

- Exchangers may be spaced apart and grouped in pairs.

- When spaced apart, a clear access way of 750 mm is considered adequate, this being the clear space between the shells or the associated pipework and insulation.

- For paired exchangers, a similar condition is required between pairs and adjacent singles, but between each shell of the pair, this may be reduced to 450 mm between head flanges.

- Exchangers always should be arranged such that a minimum of 150 mm is clear at the rear for removal of the bonnet and space is provided for dropping it clear of the working area.

- At the front or channel end, a minimum distance of the tube length plus 2500 mm is considered sufficient.

- This latter does not apply to exchangers located in structures, where a total of 1500 mm would be sufficient.

- Piping connected to heat exchangers generally should be kept simple.

- Piping economy and good engineering design depend largely on knowing what alterations can be made to exchangers. In other words, the piping designer can influence the exchanger design; for example, the direction of flow (Figure 4–2) and nozzle locations (Figure 4–3). Alterations to exchangers, of course, should not affect their duty and cost.

- The cost saved on simpler piping should not be spent on costly alterations to exchangers.

- Figure 4–4 shows the possible alterations by piping designers to typical shell and tube exchangers that do not affect the thermal design.

- When contemplating such a change, it should be remembered that generally the heated media should flow upward and the cooled media downward. This is particularly important if a physical change takes place within the exchanger, such as vaporization or condensation. Typical examples of this are:

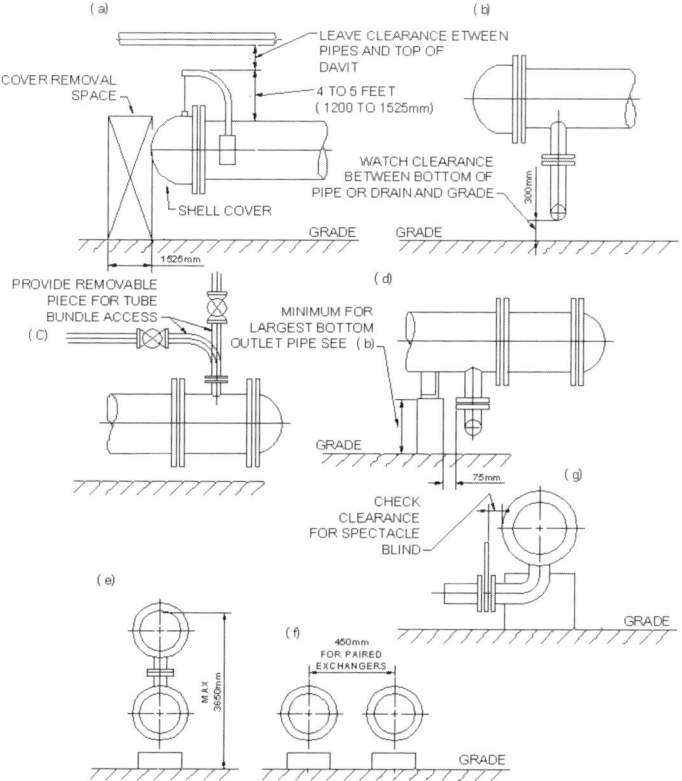

Figure 4–1   *Minimum clearances for heat exchangers (courtesy of Red Bag/Bentley Systems, Inc.).*

- Reboilers, where the process stream enters the shell at the bottom as a liquid and leaves at the top as a vapor after flowing through the tubes, and stream that enters the shell near the top of the tubes and leaves at the bottom of the shell as condensate.

- Condensers, where the process stream enters the shell at the top as a vapor and leaves the bottom as a liquid, while cooling water enters the tubes side at the bottom and leaves at the top.

Exchanger Layout Other Than in Banks

- Process equipment in most plants is arranged in the sequence of the process flow. However, whatever layout system is used, the general evaluation regarding exchanger positions is very similar.

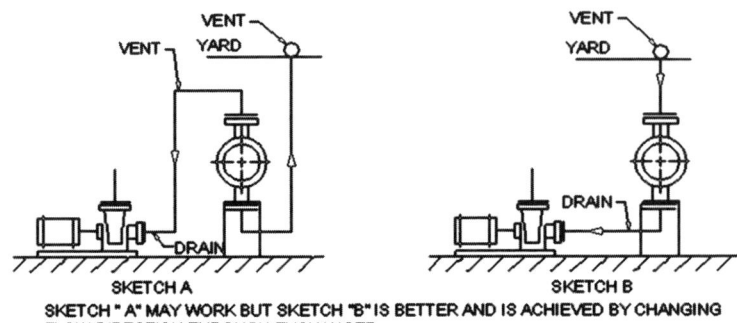

SKETCH "A" MAY WORK BUT SKETCH "B" IS BETTER AND IS ACHIEVED BY CHANGING
FLOW DIRECTION THROUGH EXCHANGER

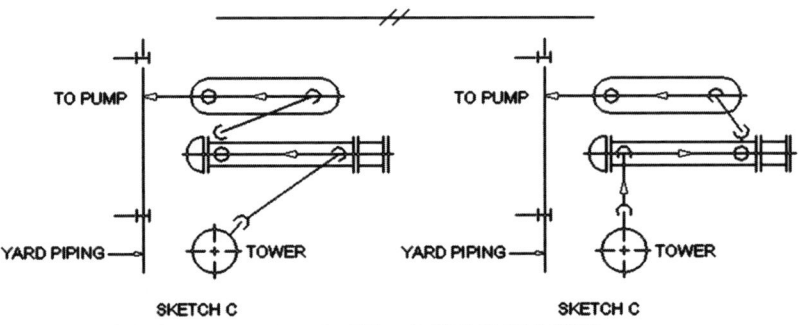

SKETCH "C" SHOWS ZIG ZAG FLOW PATTERN WHICH IS TO BE AVOIDED.
SKETCH "D" SHOWS HOW RELOCATING NOZZLE CAN PROVIDE A MORE FUNCTIONAL FLOW
PATTERN AND SHORTER PIPING.

Figure 4–2 *Better piping arrangements (courtesy of Red Bag/Bentley Systems, Inc.).*

- When laying out the plot, the fractionation towers should be located in the sequence first, although often the arrangement of other equipment (for example, condensers) depends directly on the tower orientation, and sometimes the decision whether or not to use a structure depends on this.

- The relative position of exchangers can be evaluated readily from flow diagrams. For exchanger positions in a petrochemical plant, the following general classification can be made:

  - Exchangers that must be next to other equipment. Such exchangers are the reboilers, which should be located next to their respective towers, or condensers, which should be next to their reflux drums close to the tower.

  - Exchangers that should be close to other process equipment. An example would be exchangers in closed

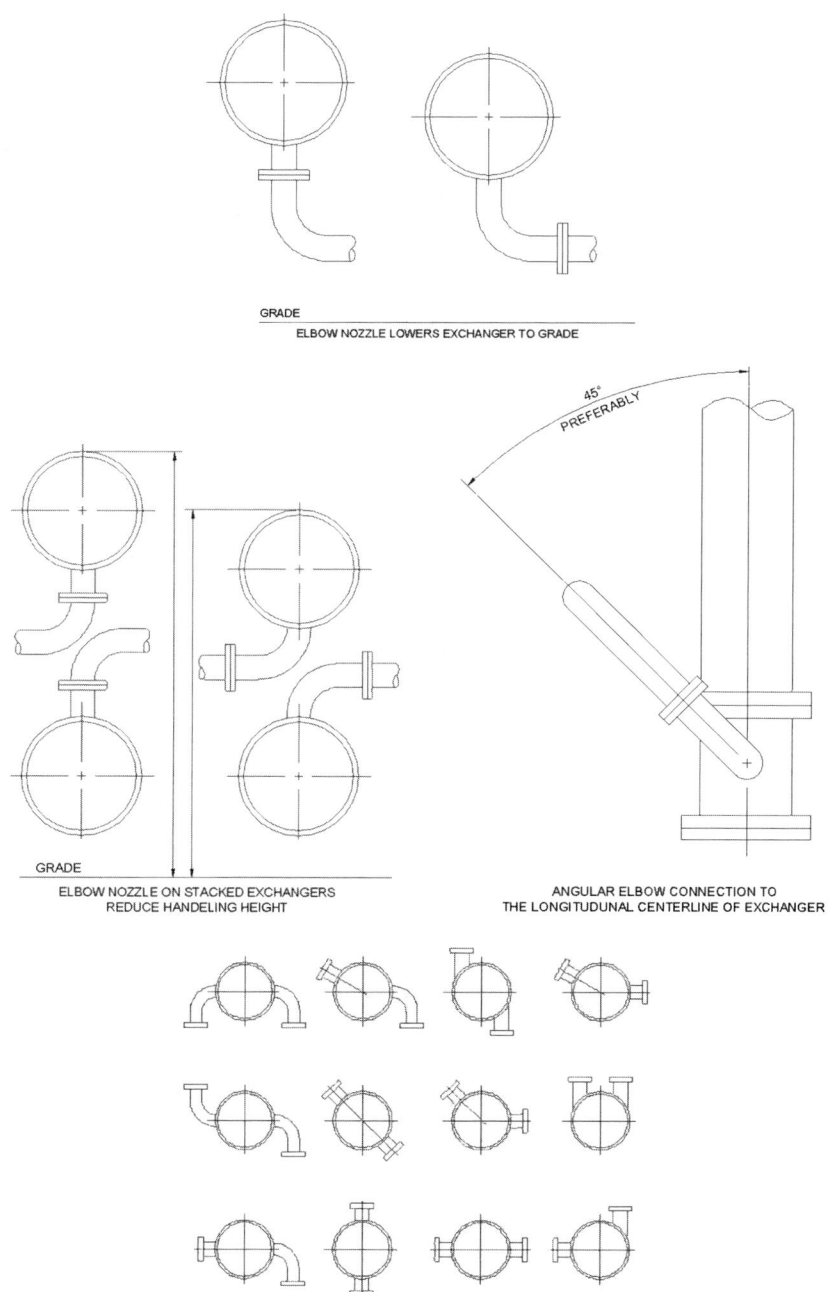

Figure 4–3   *Nozzle arrangement for better piping (courtesy of Red Bag/Bentley Systems, Inc.).*

KEY Nos.	A: TYPICAL HEAT EXCHANGE ARRANGEMENTS	B: TYPE OF TUBE BUNDLE	C: CONSTRUCTION DESCRIPTION	D: TYPICAL USE FOR TERMINOLOGY AND TYPICAL FUNCTIONS	E: POSSIBLE ALTERATIONS TO TYPICAL ARRANGEMENTS WITHOUT AFFECTING THERMAL DESIGN					
					INTERCHANGE FLOW MEDIA	CHANGE DIRECTION OF FLOW ON			CHANGE NOZZLE LOCATION	
						SHELL SIDE ONLY	TUBE SIDE ONLY	BOTH SIDES SAME TIME	TURN TUBE SIDE NOZZLE RAD. 180	TURN SHELL SIDE NOZZLE RAD. 180
1			SINGLE PASS SHELL - SINGLE PASS TUBES		✓			✓	✓	✓
2	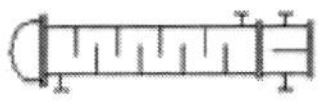		SINGLE PASS SHELL - TWO OR FOUR PASS TUBES	EXCHANGER COOLER HEATER	✓		✓	✓		✓
3		FLOATING HEAD	DOUBLE PASS SHELL - TWO OR FOUR PASS TUBES		✓			✓		
4	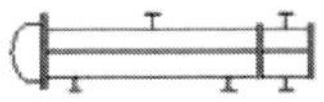		SINGLE SPLIT FLOW SHELL - TWO OR FOUR PASS TUBES	CONDENSER EVAPORATOR REBOILER			✓			
5	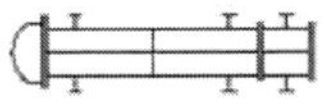		DOUBLE SPLIT FLOW SHELL - TWO OR FOUR PASS TUBES				✓			

KEY Nos.	A: TYPICAL HEAT EXCHANGE ARRANGEMENTS	B: TYPE OF TUBE BUNDLE	C: CONSTRUCTION DESCRIPTION	D: TYPICAL USE FOR TERMINOLOGY AND TYPICAL FUNCTIONS	E: POSSIBLE ALTERATIONS TO TYPICAL ARRANGEMENTS WITHOUT AFFECTING THERMAL DESIGN					
					INTERCHANGE FLOW MEDIA	CHANGE DIRECTION OF FLOW ON			CHANGE NOZZLE LOCATION	
						SHELL SIDE ONLY	TUBE SIDE ONLY	BOTH SIDES SAME TIME	TURN TUBE SIDE NOZZLE RAD. 180	TURN SHELL SIDE NOZZLE RAD. 180
6	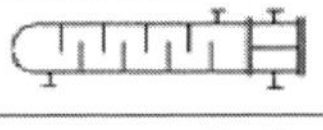		SINGLE PASS SHELL - TWO FOUR PASS TUBES	EXCHANGER COOLER HEATER	✓	✓	✓	✓		✓
7	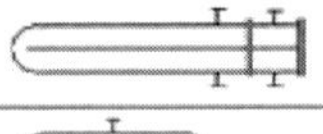	U-TUBES	DOUBLE PASS SHELL - TWO OR FOUR PASS TUBES	(IN CLEAN TUBE SIDE SERVICE)	✓					
8	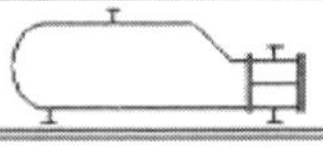		KETTLE TYPE REBOILER	REBOILER STEAM GENERATOR VAPOURIZER						
9			SINGLE PASS SHELL - SINGLE PASS TUBES	EXCHANGER COOLER HEATER	✓			✓	✓	✓
10	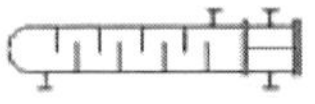	FIXED TUBES	SINGLE PASS SHELL - TWO OR FOUR PASS TUBES		✓		✓	✓		✓
11	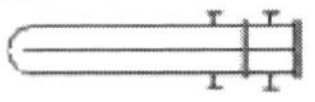		TWO PASS SHELL - TWO OR FOUR PASS TUBES	(IN LOW TEMPERATURE CLEAN SERVICE)	✓			✓		

Figure 4–4   *Typical exchangers with possible alterations for better piping (courtesy of Red Bag/Bentley Systems, Inc.).*

pump circuits, such as some reflux circuits. Overhead condensers also should be close to their tower to ensure that the line pressure drop in minimal. In case of tower-bottom-draw-off-exchanger-pump flow, exchangers should be close to the tower or drum to require short suction lines.

- Exchangers located between distant items of process equipment. An example would be exchangers with process lines connected to both shell and tube side, where parallel run is the ideal and the best location is on that side of the yard where the majority of related equipment is placed. Other locations cost more in pipe runs.

- Exchangers located between process equipment and the unit limit; for example, product coolers, which frequently are located near the unit limit.

- Stacked exchangers. A further step in the layout is to establish which exchangers can be stacked to simplify piping and save plot space. Most units in the same service are grouped automatically. Two exchangers in series or parallel usually are stacked. Sometimes, small diameter exchangers in series can be stacked three high. Two exchangers in dissimilar services also can be stacked. Sufficient clearance must be provided for shell and channel side piping between the two exchangers. Reboilers and single condensers usually stand by themselves beside their respective towers. Vertical thermosyphon reboilers usually are hung from the side of their associated tower.

Alterations to Exchangers

Various alterations to shell and tube exchangers can be considered to suit the plant layout, operational aspects, maintenance, or safety:

- Interchange flowing media between the tube and shell side. This change often is possible, more so when the flowing media are similar, for example, liquid hydrocarbons. Preferably the hotter media should flow in the tube side to avoid heat losses through the shell or the necessity for thicker insulation.

- Change direction on flow on either tube or shell side. On most exchangers in petrochemical plants, these changes frequently are possible without affecting the required duty of the exchanger, if the tubes are in a double or multipass arrangement and the shell is in a cross flow arrangement. In exchangers where counter flow conditions can be arranged, the flow direction should be changed simultaneously in the tube and shell. Some points to consider when contemplating a flow change are these:

- Shell leakage. When water cooling gases, liquid hydrocarbons. or other streams of dangerous nature. it is better to have the water in the shell and the process in the tubes, since any leakage of, say, gas will contaminate the water rather than leaking to atmosphere.

- High pressure conditions. It usually is more economical to have high pressure in the tubes than in the shell, as this allows for minimum wall thickness shell.

- Corrosion. Corrosive fluids should pass through the tubes, thus allowing the use of carbon steel for the shell.

- Fouling. It is preferable to pass the clean stream through the shell and the dirty one through the tubes. This allows for easier cleaning. Mechanical changes, such as tangential or elbowed nozzles, can assist in simplifying the piping or lowering stacked exchangers.

Elevations

- Where process requirements dictate the elevation, it usually is noted on the P&ID.

- From the economic point of view, grade is the best location of the equipment, where it also is more convenient for the tube bundle handling and general maintenance.

- Exchangers are located in structures when gravity flow to the collecting drum is required or the outlet is connected to a suction pump that has specific net positive suction head requirements. To elevate exchangers without specific requirements, the following procedure is recommended:

  - Select the exchanger with the largest bottom connection; add to the nozzle standout dimension (center line of exchanger to face of flange) the dimension through the hub of flange, elbow (1.5 diameters), one half the outside pipe diameter and 300 mm for clearance above grade.

  - Now subtract the center line to underside of support dimension from above, and the dimension remaining is the finished height of the foundation including grout.

- It is preferable if this foundation height can be made common for all the exchangers in the bank. If this is impracticable due to extremes of shell or connection pipe sizes, then perhaps two heights can be decided on.

When stacking exchangers two or three high, it is desirable that the overall height does not exceed 12' 0" (3650 mm) due to the problem of maintenance or bundle pulling.

Piping

- In the plant layout, the exchanger bank should be laid out with spacing as noted previously and all the channel nozzles on a common center line.
- This is particularly important if the cooling water (CW) headers are underground, as the CW inlets can rise into the lower channel nozzles.
- The end of the exchanger adjacent to the rack normally is the fixed end; if the CW headers are underground, the fixed end becomes the channel end.
- All changes proposed must be discussed fully with the process engineer and client engineer or Vessel Department.
- Lines turning right in the yard should be right from the exchanger center line and those turning left should approach the yard on the left-hand side of exchanger center line.
- Lines from bottom connections also should also turn up on the right or left side of exchangers, depending on which way the line turns in the yard.
- Lines with valves should turn toward the access aisle, with valves and control valves arranged close to exchanger.
- Utility lines connecting to a header in the yard can be arranged on any side of exchanger center line without increasing pipe length.
- Access to valve handwheels and instruments influence the piping arrangement around heat exchangers.
- Valve handwheels should be accessible from grade and from a convenient access way. These access ways should be utilized for arranging manifolds, control valves, and instruments.
- In the piping arrangement, provision for tube removal access should be provided. This means a spool piece of flanged elbow in the pipe line connecting to the channel nozzle.

- The requirements of good piping layout generally apply to the design of heat exchanger piping. The shortest lines and least number of fittings, temperature permitting, obviously provide the most economical piping arrangement.

- The designer should avoid loops, pockets, and crossovers and should investigate, nozzle to nozzle, the whole length of piping routed from the exchanger to some other equipment, aiming to provide no more than one high point and one low point, no matter how long the line.

- Very often a flat turn in the yard, an alternative position for control valves or manifold, changed nozzle location on the exchanger, and the like, can accomplish this requirement.

- Avoid excessive piping strains on exchanger nozzles from the actual weight of pipe and fittings and from forces of thermal expansion.

- For valves and blinds the best location is directly at the exchanger nozzle.

- In the case of an elbow nozzle on an exchanger, sufficient clearances must be provided between valve handwheel and outside of exchanger.

- Elevated valves may require a chain operation. The chain should hang freely at an accessible spot near the exchanger.

- Figures 4–5 and 4–6 show sketches highlighting exchanger piping details.

- Orifice flanges in exchanger piping usually are in horizontal pipe runs. These lines should be just above headroom and the orifice itself accessible with a mobile ladder.

- Orifices in a liquid line and mercury-type measuring element require more height.

- At gas lines the U-tube can be above the line with orifice, consequently the height is not critical.

- Lines with orifice flanges should have the necessary straight runs before and after the orifice flanges required in specification or standards.

- Locally mounted pressure and temperature indicators on exchanger nozzles, on the shell or process lines, should be visible from the access aisles.

- Similarly, gauge glasses and level controllers on exchangers should be visible and associated valves accessible from this aisles.

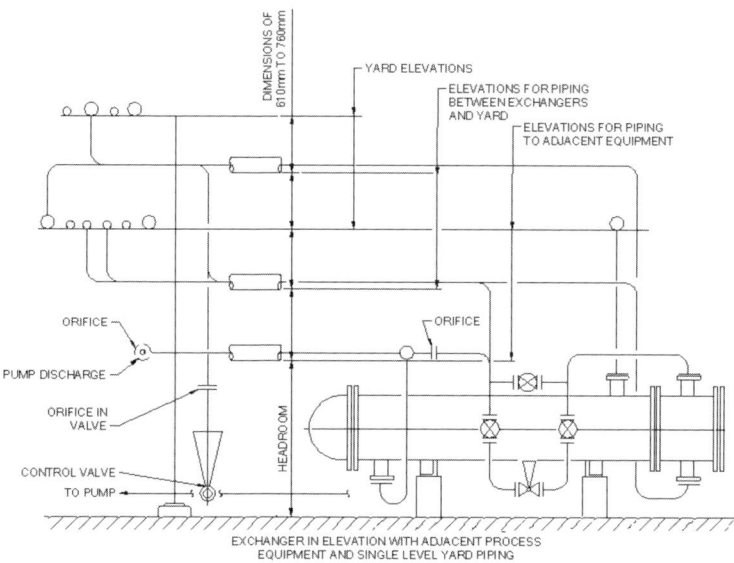

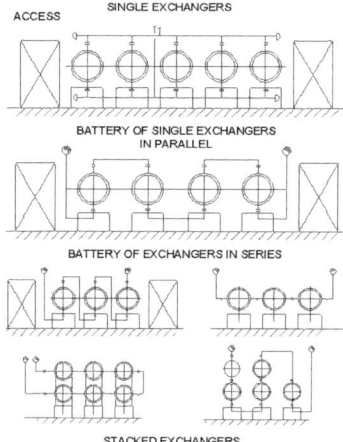

Figure 4–5 *Typical exchanger groupings (courtesy of Red Bag/Bentley Systems, Inc.).*

- When arranging instrument connections on exchangers, sufficient clearances should be left between the flanges and exchanger support between instruments and adjacent piping.
- Insulation of piping and exchangers also should be taken into account.

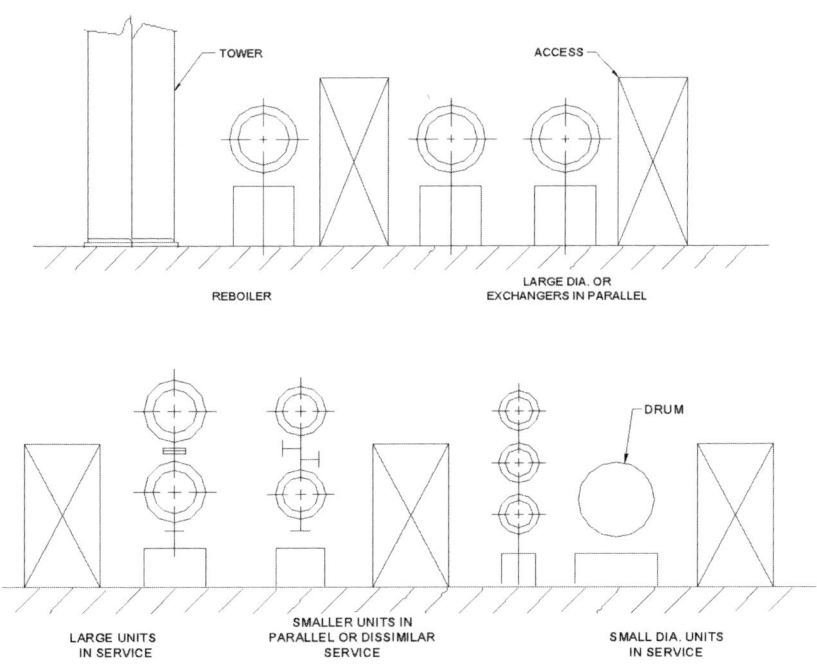

Figure 4–5   *Contintued*

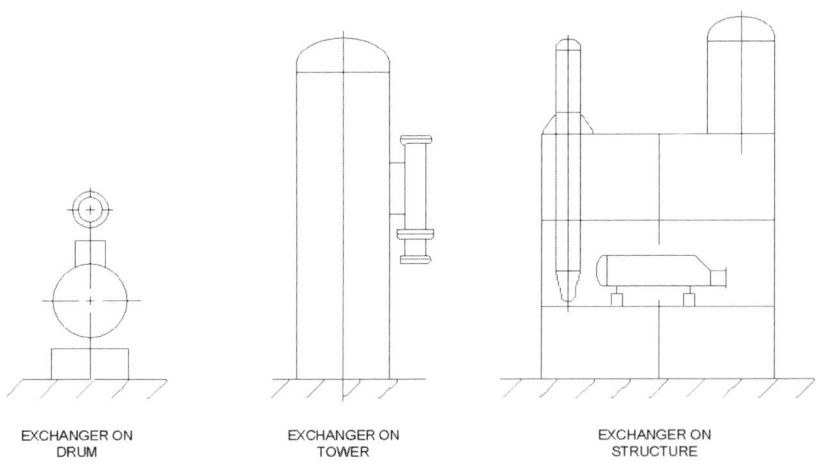

Figure 4–6   *Exchanger piping arrangement (courtesy of Red Bag/Bentley Systems, Inc.).*

4.4.2 Fin Tube Exchangers

Fin tube exchangers consist of a hairpin-shaped inner tube with heat transfer fins on the outside, except for the return bend (see Figure 4–7). The two legs are jacketed with larger bore pipe. The heat exchange is achieved by the stream passing through the hairpin and the other passing through the jackets. They may be used singly or in multiples.

The primary uses are for heaters or coolers. The process stream passes through the inner tube and either steam or cooling water passes through the jackets. They are used mainly as a source of local heat exchange, such as outlet heaters from the tanks and drums to pumps.

An important point to remember when locating fin tubes is that the hairpin tube draws out from the back end, that is, the opposite end from the nozzles, and sufficient room must be allowed for this purpose. Piping design considerations are similar to those on shell and tube exchangers.

4.4.3 Air Fins

Layout

- Air fins are large compared with shell and tube exchangers, and it is not uncommon for them to occupy several thousand square feet of plot area on a unit.
- If this plot area is required at grade, there could be plant layout problems; but fortunately most process using air fins require a gravity feed through them, which means they must be elevated.
- The most common satisfactory location is on top of the main pipe rack.
- The pipe rack width is invariably determined by tube length of air fin units.
- In the absence of sufficient room on the rack, they may be located on top of any suitable structure, or an elevated structure may be built for the purpose.

When locating air fins on the plot a number of points have to be considered:

- An air fin of a given capacity could be made up of several units, each weighing several tons. It is important that each unit be reached by the site crane for erection and maintenance purposes. Therefore, the overall plot layout must provide for this crane access (see Figure 4–8).

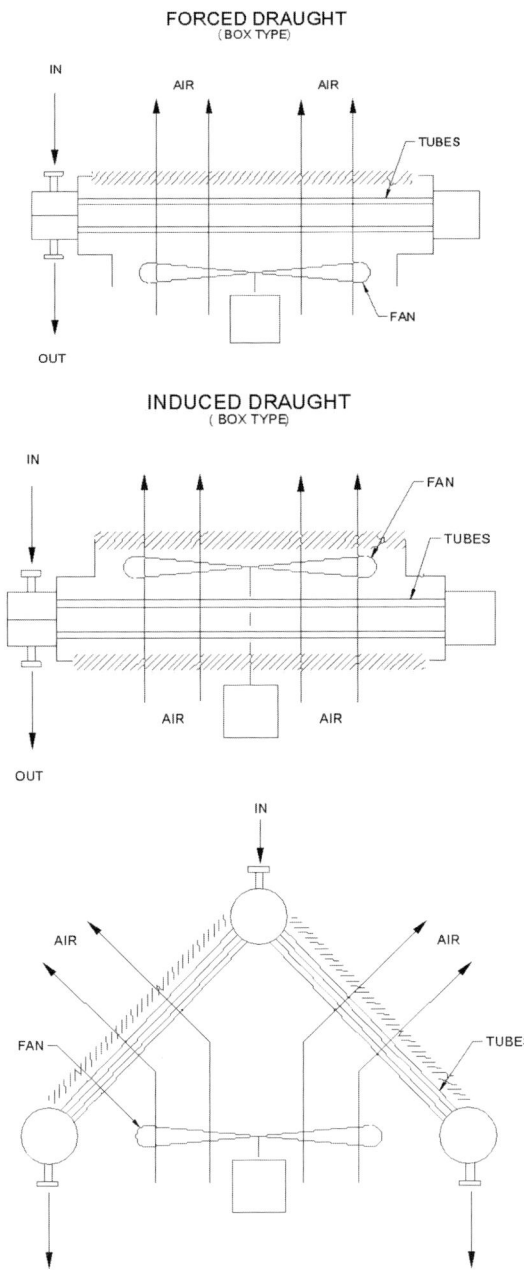

Figure 4–7   *Types of air fin exchangers (courtesy of Red Bag/Bentley Systems, Inc.).*

- As most air fins are condensing overheads from towers, it is important to consider the explanation problems of the overhead line when locating the relevant air fin, as air fins are unsuitable for accepting high loads on the nozzles.
- Access platforms always are provided either side of the air fin for access to the header boxes and underneath the units for access to fans and motors. Provision must be made for grade to all these platforms at least at either end.
- The possibility of connecting these access ways to adjacent structures to provide intermediate escape and for operational convenience.

Piping

When piping-up air fins, four major problems are encountered:

1. Correct configurations of piping to give equal or as near equal distribution as possible of the product through each unit of multiunit air fins.
2. Make piping from the tower overhead as short as possible to minimize pressure drop.
3. Obtaining a piping system that is sufficiently flexible to avoid overloading the unit nozzles.
4. Providing sufficiently suitable pipe supports and anchors.

Figure 4–9 shows diagrammatically three methods of piping for distribution: A and B show good distribution, C shows bad distribution.

- When designing the piping for air fin exchangers, the basic rules of piping still apply, that the piping runs should be as short and direct as possible but be sufficiently flexible to avoid overloading the air fin nozzles. Figure 4–10 shows two methods of running product headers to air fins.
- By running the inlet header down the center of the units, the off-takes to the unit drop out of the bottom of the header, run across the units. and drop into the nozzles. Thus, we have a series of off-takes sufficiently long to absorb expansion, at the same time, having the minimum of elbows resulting in minimum pressure drop.
- Supports usually can be attached to the steel members that run between units and therefore are short and minimal. The header must be flanged at intervals along its length to facilitate the removal of units by crane for maintenance.

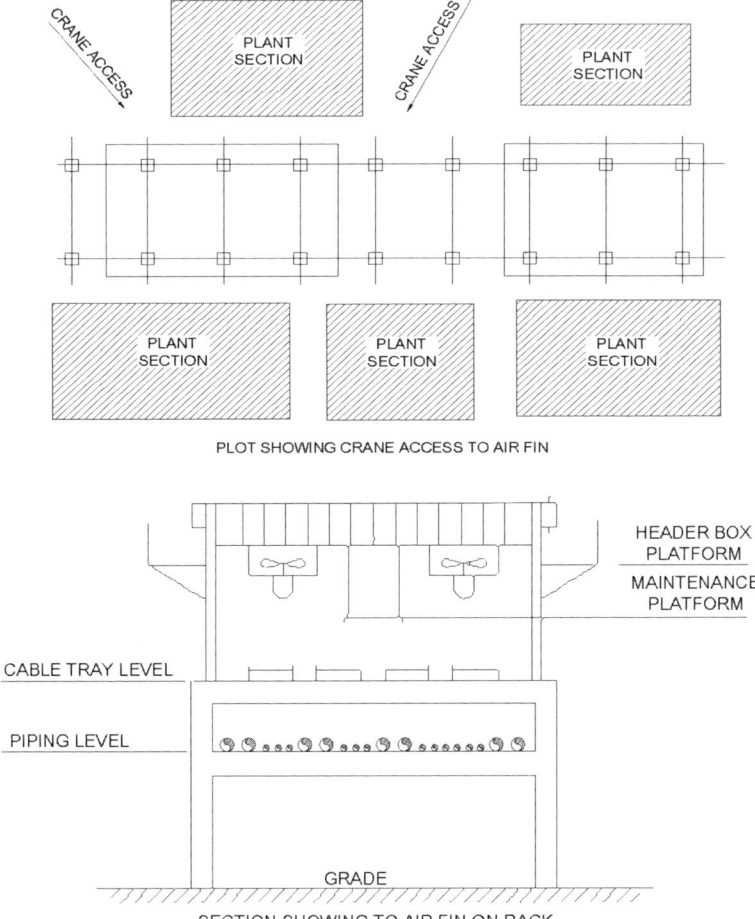

PLOT SHOWING CRANE ACCESS TO AIR FIN

SECTION SHOWING TO AIR FIN ON RACK

Figure 4–8   *Plot of crane access to air fin and section of air fin on rack (courtesy of Red Bag/Bentley Systems, Inc.).*

- The preferred position for the header is directly above the inlet nozzles, keeping the branches as short as possible. Ensure that the air fin is capable of accepting the movement imposed on the header. Support from the rack steel is between the header boxes.

- Outlet headers are less of a problem because the temperature is lower and the pipe size usually much smaller.

- They usually can be supported off the air fin legs beneath the header box platform.

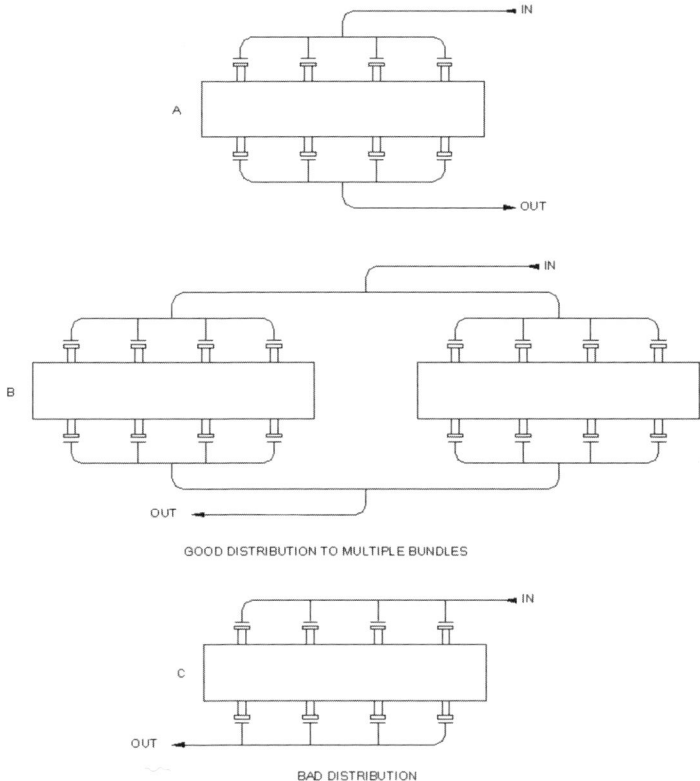

Figure 4–9 *Air fin manifold layout (courtesy of Red Bag/Bentley Systems, Inc.).*

- Any valves required to isolate units are the best located on the unit nozzles. Sometimes, air fins handling light hydrocarbons require a snuffing steam supply.
- These should be treated in the same way as snuffing steam to heaters; for example, the valves should be located at least 15 meters radius from the perimeter of the air fin.
- Piping runs that place loads of any sort on the air fin structures should be avoided if possible or communicated to the vendor as soon as possible.

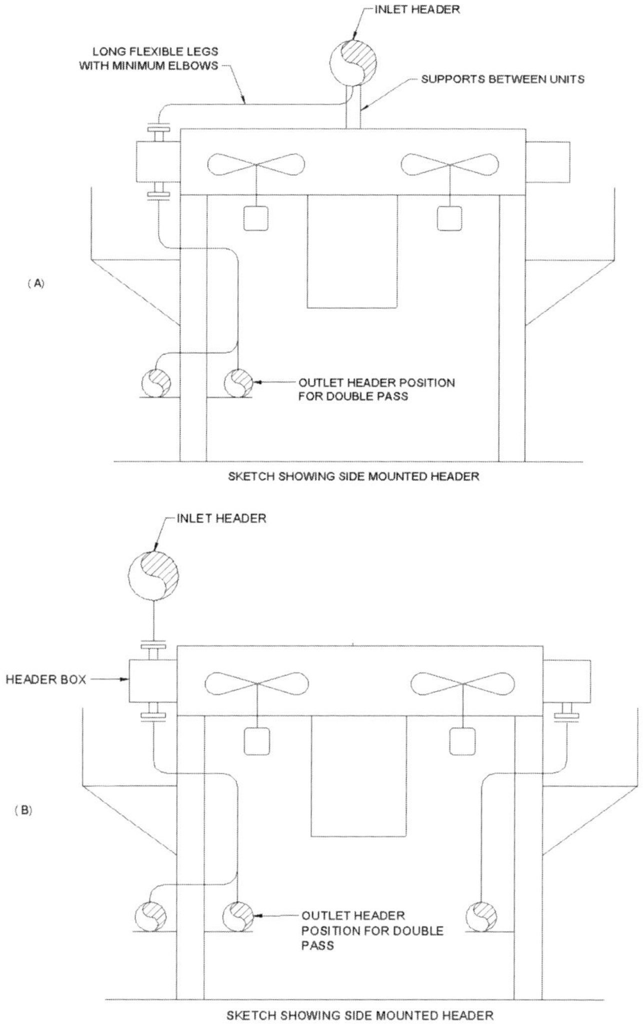

Figure 4–10   *Header mountings for air fins (courtesy of Red Bag/Bentley Systems, Inc.).*

4.5   Piping Support and Stress Issues

Shell and tube heat exchangers are relatively simple equipment when dealing with piping support or pipe stress.

- The allowable forces to these static equipment is relatively high compared with delicate rotating equipment.
- The allowable forces on the nozzles, however, are limited and should be coordinated with the mechanical or vessel engineer.
- The nozzles sometimes can be reinforced with reinforcement pads but the piping designer may be asked to reduce the forces and create more flexibility in the piping.
- Thermal stress calculations should be conducted for piping over 1.5" to the air coolers, since the allowable forces and moments are very small in comparison with the shell and tube heat exchangers.
- The manifold connected to the air coolers is a common problem for pipe stress engineers. Due to the expansion of the piping, the piping manifold pushes apart the various air cooler banks, creating extra force on the supporting structure and the piping support.

CHAPTER 5

Fired Heaters

5.1   Introduction

The primary function of a fired heater is to supply heat to the process. A fired heater utilizes gaseous or liquid fuels, often produced as a by-product from the process of the plant. The normal process function is the raising of the process fluid stream to its required temperature for distillation, catalytic reaction, and the like. Heaters vary considerably in size, depending on the type of duty and throughput.

5.2   Types of Heaters

There are two general basic designs or types of fired heaters: the box type and the vertical type. Both types of fired heater may be either forced or natural draft.

5.2.1   Box Heaters

Any heater in which the tubes are stacked horizontally is considered to be a box heater (see Figures 5–1 and 5–2). In this type of heater, it is possible to have locations or zones of different heat densities. The zone of highest heat density is called the *radiant section*, and the tubes in this section are called *radiant tubes*.

The heat pickup in the radiant tubes is mainly by direct radiation from the heating flame. In some fired heater designs, shield tubes are used between the radiant and convection sections. The zone of the lowest heat density is called the *convection section*, and the tubes in this section are called *convection tubes*. The heat pickup in the convection section is obtained from the combustion gases primarily by convection.

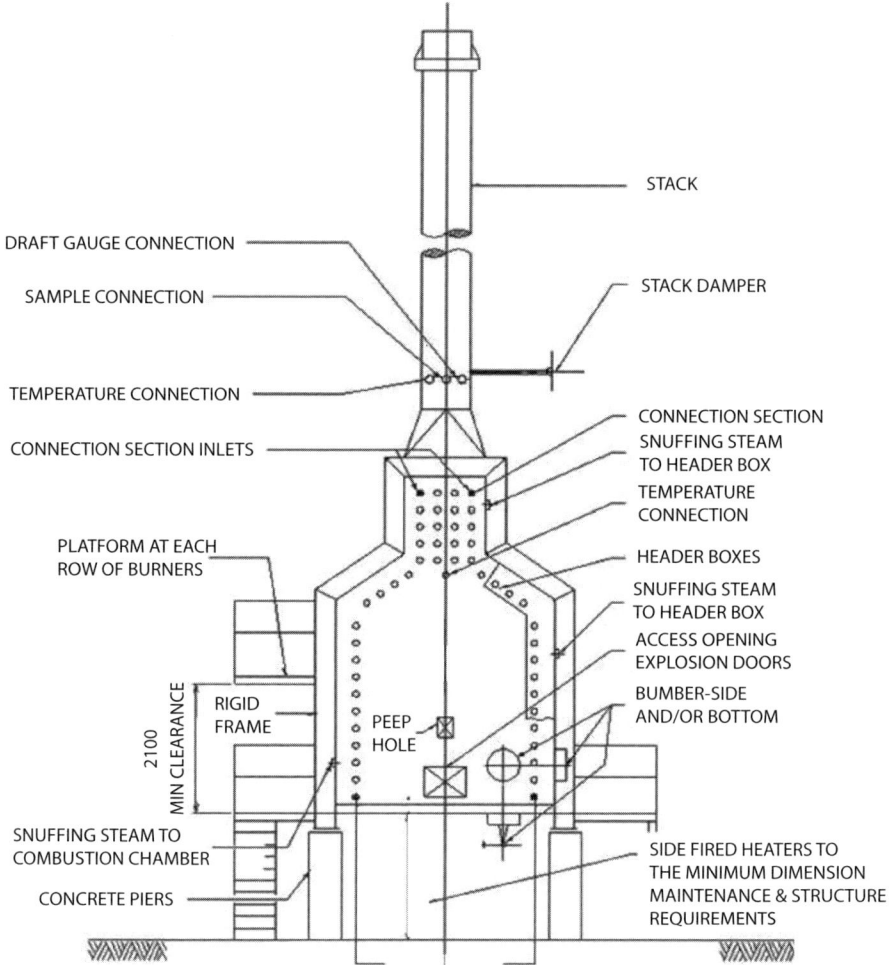

Figure 5–1   *Box heater plan (courtesy of Red Bag/Bentley Systems, Inc.).*

Box heaters may be up fired or down fired with gas or oil fired burners located in the end or sidewalls, floor, roof, or any combination these options.

Up-Fired Heaters

In the horizontal up-fired heater, products combine in the radiation chamber and pass upward through banks of roof tubes and a fire brick

Figure 5–1   *Continued*

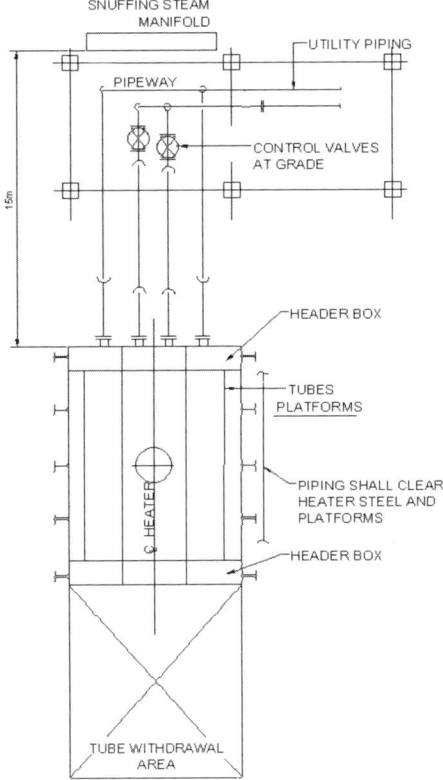

Figure 5–2   *Process piping box heater (courtesy of Red Bag/Bentley Systems, Inc.).*

diffuser into a plenum or collecting chamber. From the plenum chamber, flue gases are passed through an overhead convection section then to an overhead stack. Such heaters may be fired vertically upward by panels mounted in the heater floor or hearth, the heater floor being elevated to provide headroom.

Alternatively, these heaters may be fired horizontally by burners mounted in the heater-end walls, in which case the heater floor is elevated above grade only to provide air cooling convection to the heater foundations. This type of heater may contain single or multiple radiation chambers discharging flue gases to a common convection section and stack.

Down-Fired Heaters

In the down-fired heater, combustion gases generated in the radiant chamber pass downward through a refractory checker hearth into a collecting chamber. From there, the flue gases flow upward through the convection section then out to the stack. The down-fired heaters basically are intended to fire on heavy residual fuels, where the flue gases are corrosive and may clog flue gas passages of conventional heaters.

Convection sections are thus protected by removal of combustion solids and usually are provided with inspection ports, soot blowing devices, and tube facilities to keep the coils clean. Burners in down-fired heaters always are mounted in the heater-end walls.

5.2.2   Vertical Heaters

Vertical heaters (see Figures 5–3 through 5–5) are either cylindrical or rectangular. They may have a radiant section only or convection and radiant sections. The radiant section tubes usually are vertical, but some cylindrical heaters have helical coils. The convection section can be either vertical or horizontal.

Types of Heater Firing

Heaters can be fired from any position, bottom, top, side, or end. By far the most common is bottom fired, mainly because it is more economical. The burner of a bottom-fired heater is located 2.1–2.7 m above grade at a height suitable for an operator to work underneath. Operating from under the heater is more dangerous than other types of firing, which is the principle reason certain operating companies will not install a bottom-fired heater.

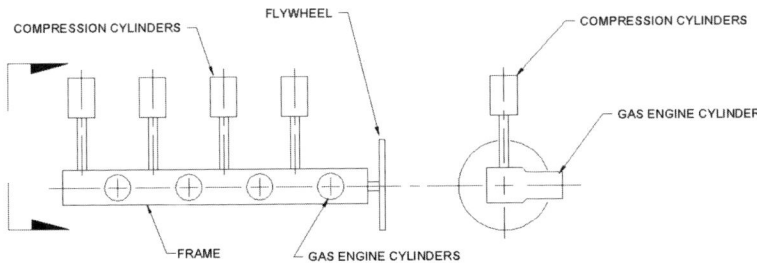

Figure 5–3 *Vertical heater with radiant convection section (courtesy of Red Bag/Bentley Systems, Inc.).*

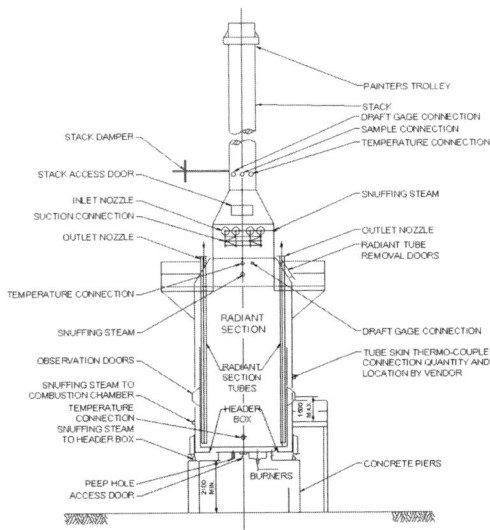

Figure 5–4 *Vertical heater with radiant section (courtesy of Red Bag/Bentley Systems, Inc.).*

Heaters are commonly lit with an electric ignite, while some refineries use a propane torch and some of the older facilities still light the burners with a rag soaked in spirits or kerosene.

Forced or Natural Draft

Consideration must be given at the layout stage to accommodate the additional equipment associated with a forced draft heater. This usually comprises an air inlet duct with silencer, forced draft fan, and an

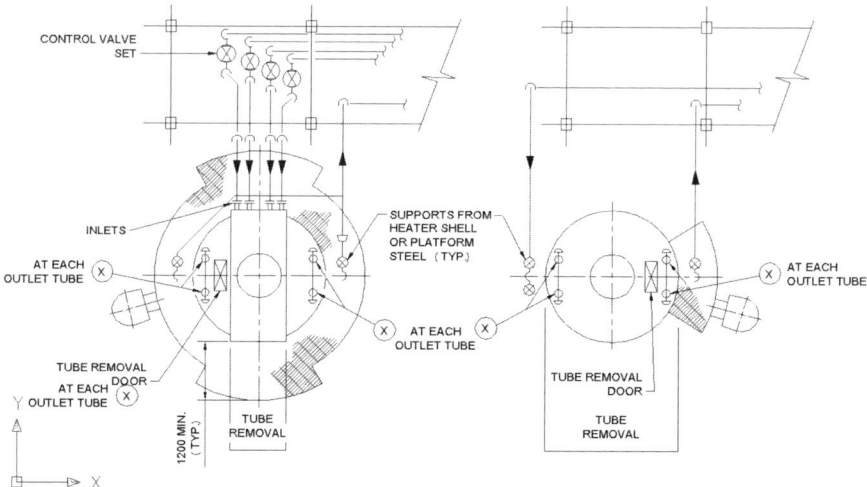

Figure 5–5   *Plan process piping at vertical heater (courtesy of Red Bag/Bentley Systems, Inc.).*

air preheater. The inlet duct may require an independent support structure.

5.3   Applicable International Codes

Numerous international codes and standards apply to fired heaters that are used in hydrocarbon processing plants and listed here are several of the most important, along with their scope and table of contents.

The design and specifying of these items of equipment are the responsibility of the mechanical engineer; however, a piping engineer or designer will benefit from being award of these documents and reviewing the appropriate sections that relate directly to piping or a mechanical-piping interface.

- API Standard 530.
- API Recommended Practice 556.
- API Standard 560.

5.3.1   API Standard 530. Calculation of Heater-Tube Thickness in Petroleum Refineries and ISO 13704:2001E), Petroleum and Natural Gas Industries—Calculation of Heater Tube Thickness in Petroleum Refineries

Scope

This international standard specifies the requirements and gives recommendations for the procedures and design criteria used for calculating the required wall thickness of new tubes for petroleum refinery heaters. These procedures are appropriate for designing tubes for service in both corrosive and noncorrosive applications. These procedures have been developed specifically for the design of refinery and related process fired heater tubes (direct-fired, heat-absorbing tubes within enclosures). These procedures are not intended to be used for the design of external piping. The international standard does not give recommendations for tube retirement thickness; annex A describes a technique for estimating the life remaining for a heater tube.

Contents

Foreword.

Introduction.

1. Scope.

2. Terms and Definitions.

3. General Design Information.

   3.1. Information Required.

   3.2. Limitations for Design Procedures.

4. Design.

   4.1. General.

   4.2. Equation for Stress.

   4.3. Elastic Design (Lower Temperatures).

   4.4. Rupture Design (Higher Temperatures).

   4.5. Intermediate temperature Range.

   4.6. Minimum Allowable Thickness

   4.7. Minimum and average Thicknesses.

   4.8. Equivalent Tube Metal Temperature.

   4.9. Return Bends and Elbows.

5. Allowable Stresses.

   5.1. General.

5.3.2 API Recommended Practice 556. Instrumentation and Control Systems for Fired Heaters and Steam Generators

Scope

his document covers recommended practices that apply specifically to instrument and control system installations for fired heaters and steam generation facilities in petroleum refineries and other hydro-carbon-processing plants. The document also discusses the installation of primary measuring instruments, control systems, alarm and shut-down systems, and automatic startup and shutdown systems for fired heaters, steam generators, carbon monoxide or waste-gas steam gener-ators, gas turbine exhaust-fired steam generators, and unfired waste heat steam generators. Although the information has been prepared primarily for petroleum refineries, much of it is applicable without change in chemical plants, gasoline plants, and similar installations.

Table of Contents

5.3.3 API Standard 560. Fired Heaters for General Refinery Service Downstream Segment

Scope

This standard covers the minimum requirements for the design, materials, fabrication, inspection, testing, preparation for shipment, and erection of fired heaters, air preheaters, fans, and burners for general refinery service. A fired heater is an exchanger that transfers heat from the combustion of fuel to fluids contained in tubular coils within an internally insulated enclosure.

Table of Contents

11.5. Fan Housing Connections.

11.6. External Forces and Moments.

11.7. Rotating Elements.

11.8. Shaft Sealing of Fans.

11.9. Critical Speeds of Resonance.

11.10. Vibration and Balancing.

11.11. Bearings and Bearing Housings.

11.12. Lubrication.

11.13. Materials.

11.14. Welding.

11.15. Nameplates and Rotation Arrows.

11.16. Drivers and Accessories.

11.17. Couplings and Guards.

11.18. Controls and Instrumentation.

11.19. Dampers or Variable Inlet Vanes.

11.20. Piping and Appurtenances.

11.21. Coatings, Insulation, and Jacketing.

11.22. Inspection and Testing.

11.23. Preparation for Shipment.

11.24. Vendor Data.

12. Instrument and Auxiliary Connections.

12.1. Flue Gas and Air.

12.2. Process Fluid Temperature.

12.3. Auxiliary Connections.

12.4. Tube-Skin Thermocouples.

13. Shop Fabrication and Field Erection.

13.1. General.

13.2. Steel Fabrication.

13.3. Coil Fabrication.

13.4. Painting and Galvanizing.

13.5. Refractories and Insulation.

13.6. Preparation for Shipment.

13.7. Field Erection.

14. Inspection, Examination, and Testing: Weld Inspection and Examination.

14.1. General.

14.2. Weld Inspection and Examination.

14.3. Castings Examination.

14.4. Examination of Other Components.

14.5. Testing.

Appendix A. Equipment Data Sheets.

Appendix B. Purchaser's Checklist.

Appendix C. Proposed Shop Assembly Conditions.

Appendix D. Stress Curves for Use in the Design of Tube Support Elements.

Appendix E. Air Preheat Systems for Fired Process Heaters.

Flue Gas Dew Point Discussion and Bibliography.

Appendix F. Measurement of the Efficiency of Fired Process Heaters.

Appendix F.A. Model Format for Data Sheets.

Appendix F.B Model Format for Work Sheets Natural Draft.

5.4   Piping—Specific Guidelines to Layout

5.4.1   Location

- Always locate heaters at a safe distance 15 m away from other hydrocarbon-bearing equipment and preferably upwind; however, in some process plants, reactors may be within this distance to prevent light volatile vapors from being blown toward an open flame.

- Space must be allowed for tube replacement for both horizontal and vertical heaters, and together with ample access for mobile equipment, this should be considered on piping layouts and drawings.

- Ample access always is needed for firefighting equipment and personnel, with areas under or around heaters usually paved and curbed.

- No low points in the paving or grading are permitted, as these provide excellent locations for trapping hydrocarbon liquids, which could be ignited by the open flames of the burners.

5.4.2 Safety

Snuffing Steam Station

- Snuffing steam connections are supplied by the heater manufacturers, generally in the combustion chamber and header boxes, see Figure 5–6.
- The control point or snuffing steam manifold generally is located at least 15 m away from the heater, is supplied by a live steam header, and is ready for instantaneous use.
- Smothering lines should be free from low pockets and so arranged as to have all drains grouped near the manifold.
- Collected condensate (apart from freezing and blocking lines), when blown into a hot furnace, can result in serious damage. (Low points should be drilled 10 mm diameter or provided with spring opening autodrain valves.)

Monitor Nozzles

- Some customers request that turret or monitor nozzles be located around the heater so that water for fighting a fire is available instantly.
- Controls for such nozzles should be located at least 15 m away from the heater.

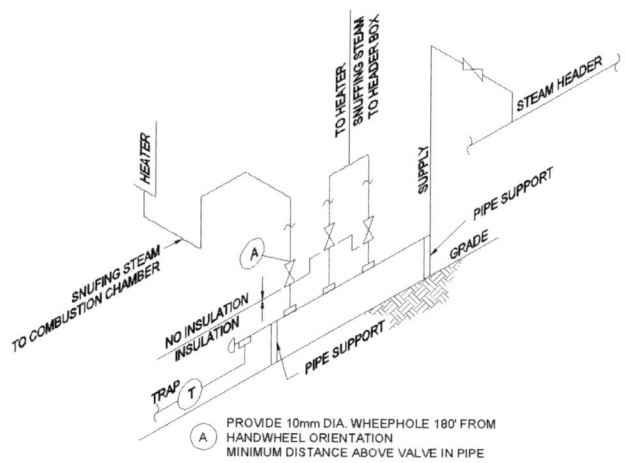

Figure 5–6 *Snuffing steam station (courtesy of Red Bag/Bentley Systems, Inc.).*

Utility Stations

- Steam, air, service water, and nitrogen connections should be provided near the tube ends of heaters.

- Process control stations, process feed and discharging block valves, and flow control valves should be located at a distance from the heater, if indicated as necessary by the engineering flow diagram or instructed by the piping specialist engineer or Project Department.

Explosion Doors

Piping should not obstruct explosion doors or tube-access doors.

Inlet and Outlet Piping

- Flow distribution through a multipass heater is affected by the inlet and outlet manifolds. The piping arrangement at the inlet of the furnace is the more critical and requires careful attention, see Figures 5–7 and 5–8.

- Inlet piping is preferred to be symmetrical and of the same length from the point where flow splits to the heater inlets. This refers to the number of bends, elbows, and valve, as well as the number of straight runs of pipe and their location.

- Piping should avoid dead-end tee branches and sharp turns.

- Unequal flow through any part of the heater results in deposits of coke and overheating of the tube walls.

- On outlet piping, symmetry is not as critical as on the inlet piping; however, nonsymmetrical piping may contribute to possible coking and overheating of tubes.

- A nonreturn and a shutoff valve usually are located at the outlet of the furnace to eliminate any flow reverse in the case of tube failure.

Burner Piping

- Supply of fuel to individual burners is adjusted by individual valves. These should be so located that the burners can be operated while observing the flame through peepholes or burner openings, see Figures 5–9 and 5–10.

- All burner leads for gas and steam (atomizing) must be taken from the top of the headers, and fuel gas piping should be so arranged that there are no pockets in which condensate could collect.

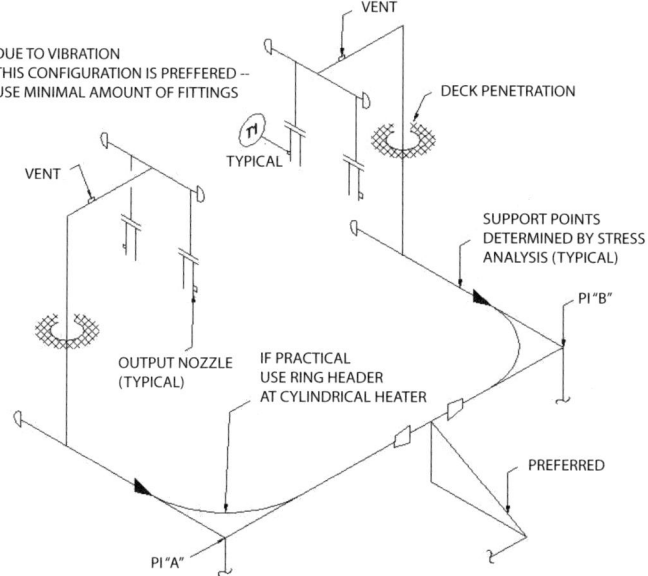

SYMMETRICAL PIPING HEADER OUTLETS (4-PASS)

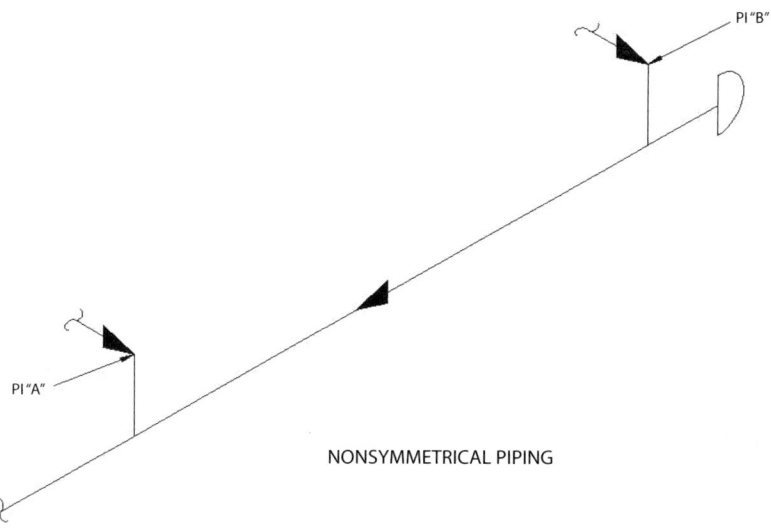

NONSYMMETRICAL PIPING

Figure 5–7   *Inlet and outlet piping (courtesy of Red Bag/Bentley Systems, Inc.).*

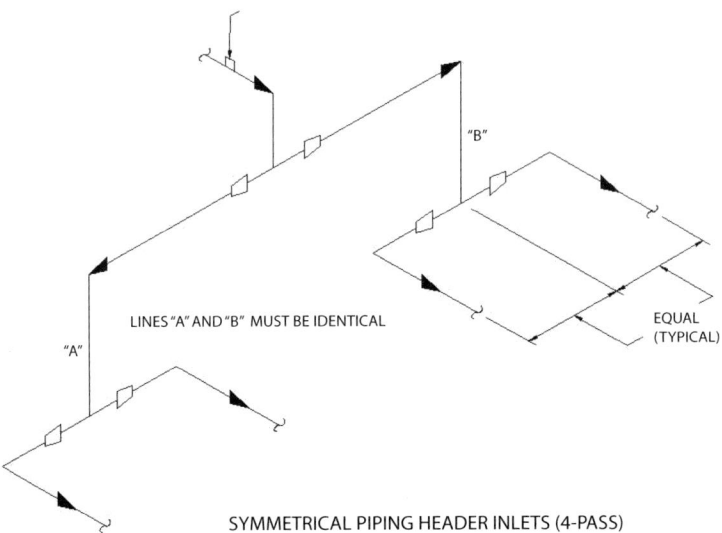

LINES "A" AND "B" MUST BE IDENTICAL

"A"

"B"

EQUAL
(TYPICAL)

SYMMETRICAL PIPING HEADER INLETS (4-PASS)

Figure 5–7   *Continued*

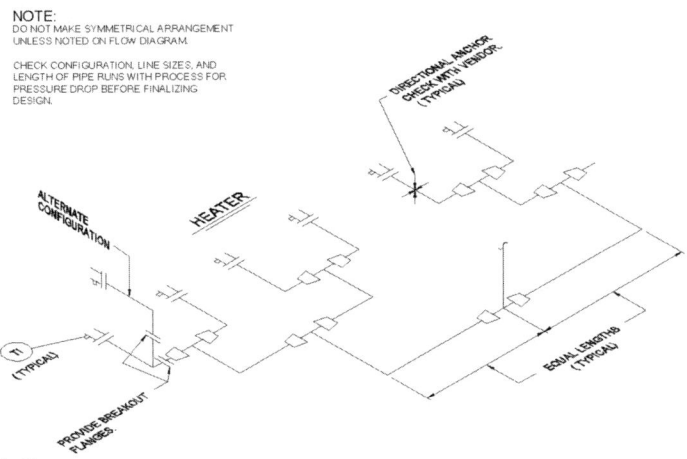

Figure 5–8   *Piping manifold to heater (courtesy of Red Bag/Bentley Systems, Inc.).*

- The fuel oil header must have full circulation, and under no circumstances should it be a dead-end line. Noncirculating branches to burners should be as short as possible or insulated together with atomizing steam.

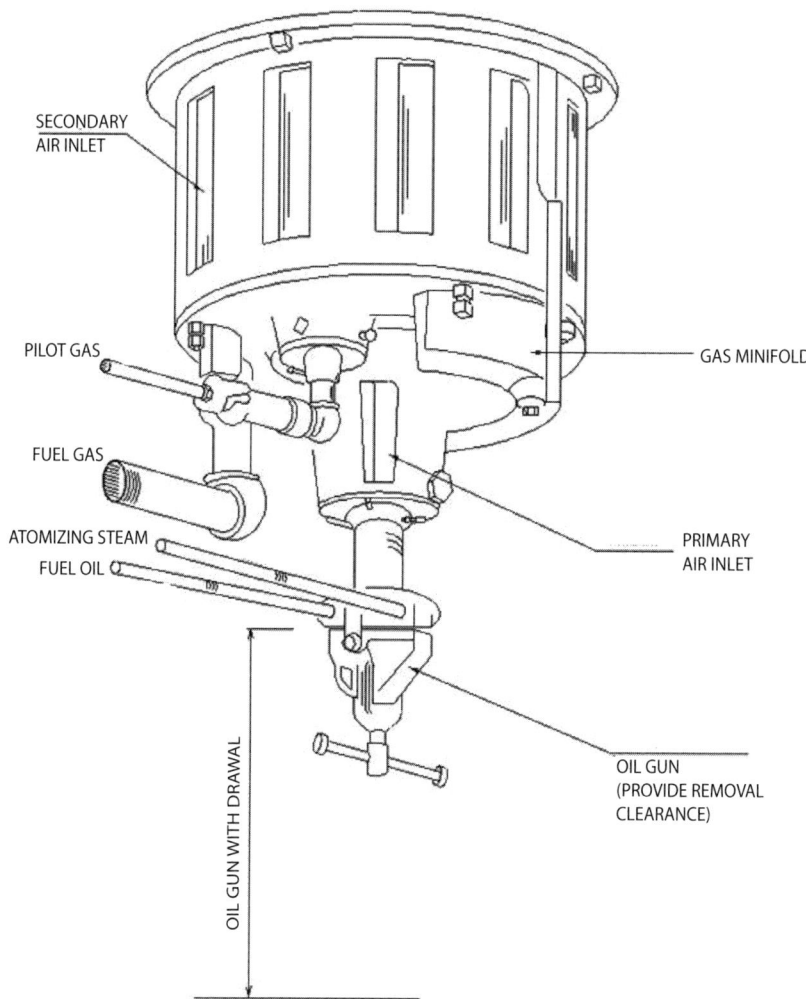

SECONDARY
AIR INLET

PILOT GAS

GAS MINIFOLD

FUEL GAS

ATOMIZING STEAM
FUEL OIL

PRIMARY
AIR INLET

OIL GUN WITH DRAWAL

OIL GUN
(PROVIDE REMOVAL
CLEARANCE)

Figure 5–9 *Natural draft burner for fuel oil and fuel gas (courtesy of Red Bag/Bentley Systems, Inc.).*

- A ring header around the furnace, mounted a short distance above the peepholes and having vertical leads adjacent to the vertical doors to the burners, provides the greatest degree of visibility from the operator's point of view.

- Atomizing steam to be used in conjunction with fuel oil should be taken from the main steam supply header at or near the heater.

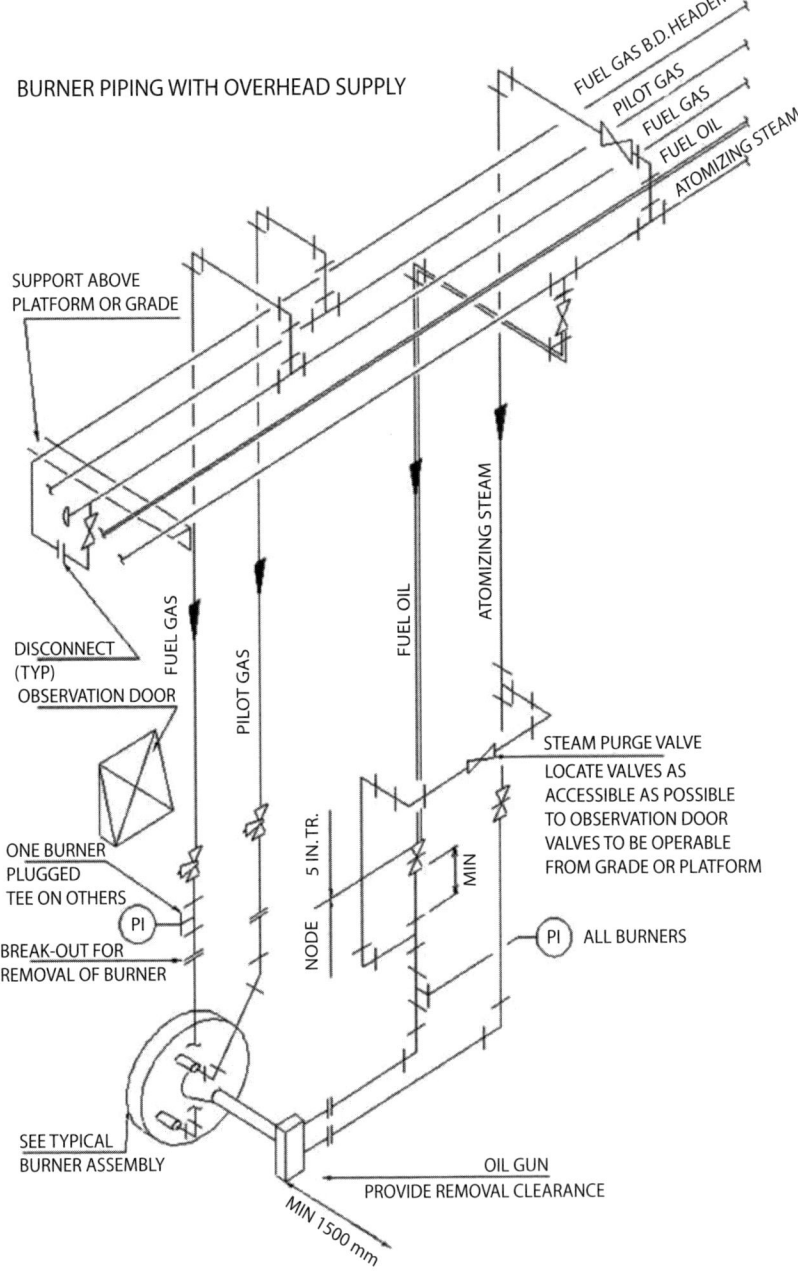

Figure 5–10   *Burner piping (courtesy of Red Bag/Bentley Systems, Inc.).*

- Steam traps should be provided to drain all low points in the atomizing steam system.
- Separate leads to each burner should be taken from the top of the atomizing steam subheader.
- Shutoff valves should be so located that they can be operated while observing the flames from the observations ports.

5.5 Operation and Maintenance Influencing Piping Design

Fired heaters require specific operation and maintenance. The operation and maintenance is determined by the manufacturer, and this needs to be considered during the piping layout and design phase.

5.5.1 Decoking

The internal cleaning of tubes and fittings may be accomplished by several methods. One is to circulate gas oil through the coil after the heater has been shutdown but before the coils are steamed and water washed and prior to the opening and start of inspection work, see Figure 5–11.

This method is effective if deposits in the coil are such that they are softened or dissolved by gas oil. When tubes are coked or contain hard deposit, other methods may be used, such as steam air decoking and mechanical cleaning for coke deposits and chemical cleaning for salt deposits.

Chemical cleaning and steam air decoking are preferable, as they tend to clean the tube to bare metal. The chemical cleaning process requires circulation of an inhibited acid through the coil until all deposits have been softened and removed. This usually is followed by water washing to flush all deposits from the coil. The steam air decoking process uses steam, air, and heat to remove the coke. The mechanics of decoking are

- Shrinking and cracking the coke loose by heating tubes from outside while steam blows coke from the coil.
- Chemical reaction of hot coke with steam.
- Chemical reaction of coke and oxygen in air.

Steam and air services are connected permanently to the heater. The heater outlet line incorporates a swing elbow that, during the decoking operation, is disconnected from the outlet line and connected to the decoking header. Care must be taken to allow sufficient

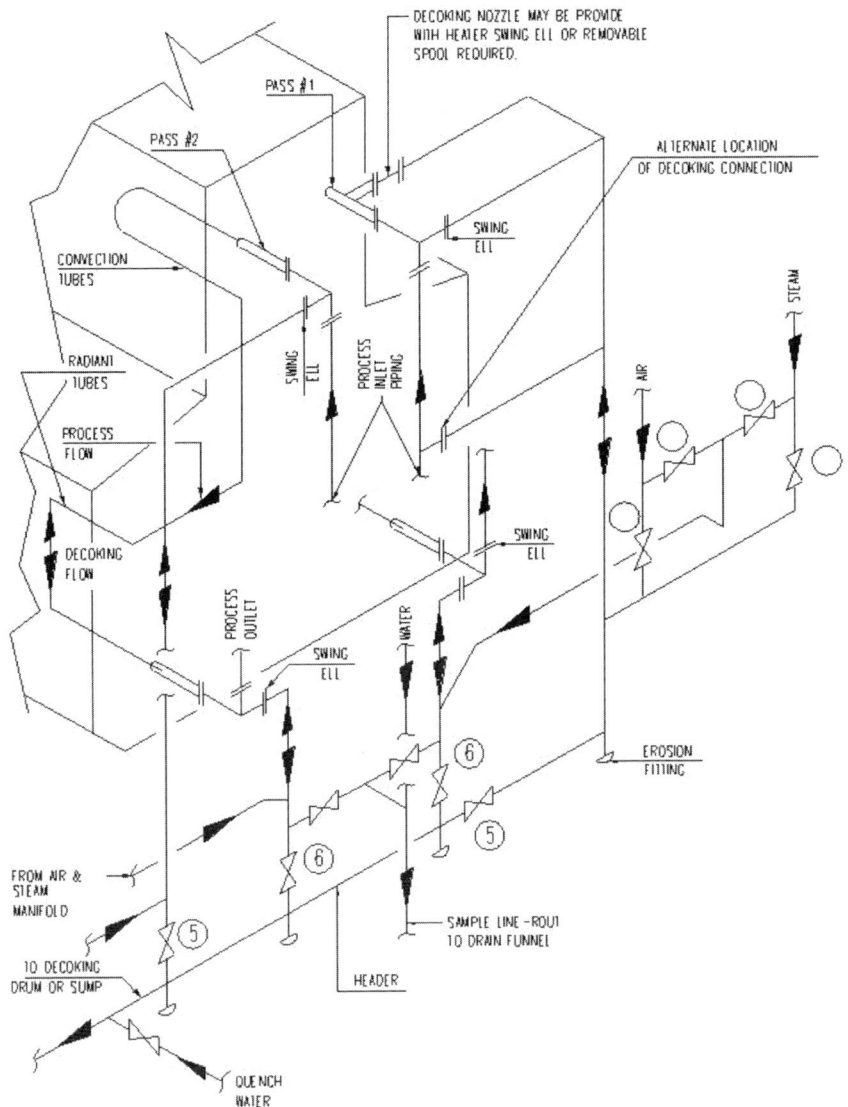

Figure 5–11   *Piping and decoking of the heater (courtesy of Red Bag/Bentley Systems, Inc.). After tubes become headed, steam is injected at the convection inlets, valves 1, 2, 4, and 5 are closed; 3 is open. To start burning, steam flow is reduced and air is introduced by opening valve 4. For reverse flow, valves 2, 3, 4, and 6 are closed; 1 and 5 are open. Valve 2 is opened only if reverse burning is required. While pass 1 is being decoked, steam is injected into pass 2 to keep the tubes cool.*

access and platform space when the swing elbows are changed over. Coke is carried by this header to the drum or sump.

In some instances, the Process Department or a client may request that the decoking manifold be connected to allow for reverse flow during the decoking.

5.5.2 Soot Blowing

In some heaters, the convection section contains tubes with extended surfaces in the form of either fins or studs. Extended-surface tubes are used to increase the convection heat transfer area at low capital cost. Because of the tendency of extended-surface tubes to foul when burning heavy oils, soot blowers usually are installed.

Soot blowers employ high-pressure steam to clean the tube outer surfaces of soot and other foreign material. Soot blowers may be either automatic electric motors operated by a pushbutton at grade or manual instruments requiring operation from a platform located at the convection bank level. Care must be taken that sufficient clearance is allowed for the withdrawal of soot blowers. Generally, heaters are supplied with soot blowing facilities in the convection section, although tubes may not be of the extended-surface type.

5.5.3 Instrument Accessibility

Various instruments are required to control and monitor the healthy operation of the heater. Both the stack and the heater body contain instrumentation.

Heater Stack

- Damper, mechanical or pneumatically operated to control the draft through the stack.
- Draft gauge (pressure gauge and pressure indicator).
- Flue temperature (temperature indicator).
- Orsat (O_2 CO CO_2 analyzer).

An instrument draft gauge, flue temperature, and Orsat are used to access the correct combustion conditions. Steam is supplied for the Orsat connection in the stack. Water is supplied to the O_2 analyzer. A platform for the access to the stack instruments is supplied with the heater.

Heater Body

- Skin thermocouples or tube wall TIs, to indicate overheating of tubes.
- PIs, to measure the draft pressure through the combustion section of the heater.

5.6   Piping Support and Stress Issues

Because of the high temperature involved and length of pipe runs required to isolate the heaters, piping flexibility must be examined carefully. Some heater manufacturers permit a limited amount of tube movement to take all or part of the piping expansion. This possibility should be investigated in conjunction with the Stress Department, by the Piping Design Office, during the study stage. It generally is necessary to anchor the piping adjacent to the heater to remove stresses from the nozzles.

CHAPTER 6

Tanks

6.1   Introduction

Tanks are used to store liquid, gas, and vapor for the operation of a process plant. The location of such equipment usually is outside the process area. The area where the tanks are located is called a *tank farm*. The material stored in the tanks can be

- Raw material to feed a process unit.
- Intermediate products.
- End products.
- Utilities such as fuel for fired heaters.
- Waste products.
- Rainwater.
- Fire water.

The construction materials of all tanks must be compatible with media that is contained within the walls. The external environmental conditions also must be considered when specifying the materials of construction of a tank and its external coating.

6.2   Types of Tanks

Basically, there are two types of tanks: atmospheric tanks and pressurized tanks. The atmospheric tanks are used for fluids that can be stored under ambient conditions, such as crude oil, water, and certain gases. Pressurized tanks are used for liquids that need to be stabilized and stored under pressure such as LPG. (liquefied petroleum gas), LNG (liquefied natural gas), and liquid nitrogen.

The atmospheric tanks can have a floating roof, which, as described, floats on the fluid or a fixed cone roof, which is attached to the walls of the tank. The floating roof is selected to reduce the vapor area above the liquid and therefore minimize the fire hazard

The pressurized tanks types can be of a spherical or bullet shape, and these storage tanks are used to store of fluids at pressures above atmospheric.

6.3   Applicable International Codes

Numerous international codes and standards apply to the various types of tanks and storage vessels used in hydrocarbon processing plants, and listed here are several of the most important along with their scope and table of contents. The design and specifying of these items of equipment are the responsibility of the mechanical engineer; however, a piping engineer or designer will benefit from being award of these documents and reviewing the appropriate sections that relate directly to piping or a mechanical-piping interface.

- API Recommended Practice 575.
- API Standard 620.
- API Standard 650.
- API Standard 653.

6.3.1   API Recommended Practice 575. Guidelines and Methods for Inspection of Existing Atmospheric and Low-Pressure Storage Tanks

Scope

This document provides useful information and recommended practices for the maintenance and inspection of atmospheric and low-pressure storage tanks. While some of these guidelines may apply to other types of tanks, these practices are intended primarily for existing tanks constructed to API Specification 12A or 12C and API Standard 620 or 650. This recommended practice includes

- Descriptions of the various types of storage tanks.
- Construction standards.
- Maintenance practices.
- Reasons for inspection.
- Causes of deterioration.
- Frequency of inspection.

- Methods of inspection.
- Inspection of repairs.
- Preparation of records and reports.
- Safe and efficient operation.
- Leak prevention methods.

This recommended practice is intended to supplement API Standard 653, which provides the minimum requirements for maintaining the integrity of storage tanks after they have been placed in service.

Table of Contents

6.3.2 API Standard 620. Design and Construction of Large, Welded, Low-Pressure Storage Tanks

Contents

4. Materials.
    4.1. General.
    4.2. Plates.
    4.3. Pipe, Flanges, Forging, and Castings.
    4.4. Bolting Material.
    4.5. Structural Shapes.
5. Design.
    5.1. General.
    5.2. Operating Temperature.
    5.3. Pressures Used in Design.
    5.4. Loadings.
    5.5. Maximum Allowable Stress for Walls.
    5.6. Maximum Allowable Stress Values for Structural Members and Bolts.
    5.7. Corrosion Allowance.
    5.8. Linings.
    5.9. Procedure for Designing Tank Walls.
    5.10. Design of Sidewalls, Roofs, and Bottoms.
    5.11. Special Considerations Applicable to Bottoms That Rest Directly on Foundations.
    5.12. Design of Roof and Bottom Knuckle Regions and Compression-Ring Girders.
    5.13. Design of Internal and External Structural Members.
    5.14. Shapes, Locations, and Maximum Sizes of Wall Openings.
    5.15. Inspection Openings.
    5.16. Reinforcement of Single Openings.
    5.17. Reinforcement of Multiple Openings.
    5.18. Design of Large, Centrally Located, Circular Openings in Roofs and Bottoms.
    5.19. Nozzle Necks and Their Attachments to the Tank.
    5.20. Bolted Flanged Connections.
    5.21. Cover Plates
    5.22. Permitted Types of Joints.
    5.23. Welded Joint Efficiency.
    5.24. Plug Welds and Slot Welds.
    5.25. Stress Relieving.

7.15. Examination Method and Acceptance Criteria.

7.16. Inspection of Weld.

7.17. Radiographic/Ultrasonic Examination Requirements.

7.18. Standard Hydrostatic and Pneumatic Tests.

7.19. Proof Tests for Establishing Allowable Working Pressures.

7.20. Test Gauges.

8. Marking.

8.1. Nameplates.

8.2. Division of Responsibility.

8.3. Manufacturer's Report and Certificate.

8.4. Multiple Assemblies.

9. Pressure- and Vacuum-Relieving Devices.

9.1. Scope.

9.2. Pressure Limits.

9.3. Construction of Devices.

9.4. Means of Venting.

9.5. Liquid Relief Valves.

9.6. Marking.

9.7. Pressure Setting of Safety Devices.

Appendix A. Technical Inquiry Responses.

Appendix B. Use of Materials That Are Not Identified with Listed Specifications.

Appendix C. Suggested Practice Regarding Foundations.

Appendix D. Suggested Practice Regarding Supporting Structures.

Appendix E. Suggested Practice Regarding Attached Structures (Internal and External).

Appendix F. Examples Illustrating Application of Rules to Various Design Problems.

Appendix G. Considerations Regarding Corrosion Allowance and Hydrogen-Induced Cracking.

Appendix H. Recommended Practice for Use of Preheat, Post-Heat, and Stress Relief.

Appendix I. Suggested Practice for Peening.

Appendix J. (Reserved for Future Use).

Appendix K. Suggested Practice for Determining the Relieving Capacity Required.

Appendix L. Seismic Design of Storage Tanks.

Appendix M. Recommended Scope of the Manufacturer's Report.

Appendix N. Installation of Pressure-Relieving Devices.

Appendix O. Suggested Practice Regarding Installation of Low-Pressure Storage Tanks.

Appendix P. NDE and Testing Requirements Summary Hydrocarbon Gases.

Appendix Q. Low-Pressure Storage Tanks for Liquefied Hydrocarbon Gases.

Appendix R. Low-Pressure Storage Tanks for Refrigerated Products.

Appendix S. Austenitic Stainless Steel Storage Tanks.

Appendix U. Ultrasonic Examination in Lieu of Radiography.

6.3.3   API Standard 650. Welded Steel Tanks for Oil Storage

Scope

This standard covers the material, design, fabrication, erection, and testing requirements for vertical, cylindrical, aboveground, closed- and open-top, welded steel storage tanks in various sizes and capacities for internal pressures approximating atmospheric pressure (internal pressures not exceeding the weight of the roof plates), but a higher internal pressure is permitted when additional requirements are met. This standard applies only to tanks whose entire bottom is uniformly supported and tanks in nonrefrigerated service that have a maximum design temperature of 93°C (200°F) or less.

The standard is designed to provide the petroleum industry with tanks of adequate safety and reasonable economy for use in the storage of petroleum, petroleum products, and other liquid products commonly handled and stored by the various branches of the industry. This standard does not present or establish a fixed series of allowable tank sizes; instead, it is intended to permit the purchaser to select whatever size tank may best meet its needs. This standard is intended to help purchasers and manufacturers in ordering, fabricating, and erecting tanks; it is not intended to prohibit purchasers and manufacturers from purchasing or fabricating tanks that meet specifications other than those contained in this standard.

Table of Contents

Appendix P. Allowable External Loads on Tank Shell Openings.

Appendix R. Load Combinations.

Appendix S. Austenitic Stainless Steel Storage Tanks.

Appendix T. NDE Requirements Summary.

Appendix U. Ultrasonic Examination in Lieu of Radiography.

Appendix V. Design of Storage Tanks for External Pressure.

6.3.4 API Standard 653. Tank Inspection, Repair, Alteration, and Reconstruction

Scope

This standard covers steel storage tanks built to API Standard 650 and its predecessor, API 12C. It provides minimum requirements for maintaining the integrity of such tanks after they have been placed in service and addresses inspection, repair, alteration, relocation, and reconstruction. The scope is limited to the tank foundation, bottom, shell, structure, roof, attached appurtenances, and nozzles to the face of the first flange, first threaded joint, or first welding-end connection. Many of the design, welding, examination, and material requirements of API Standard 650 can be applied in the maintenance inspection, rating, repair, and alteration of in-service tanks. In the case of apparent conflicts between the requirements of this standard and API Standard 650 or its predecessor, API 12C, this standard governs for tanks that have been placed in service.

The standard employs the principles of API Standard 650; however, storage tank owners and operators, based on consideration of specific construction and operating details, may apply this standard to any steel tank constructed in accordance with a tank specification.

The standard is intended for use by organizations that maintain or have access to engineering and inspection personnel technically trained and experienced in tank design, fabrication, repair, construction, and inspection.

This standard does not contain rules or guidelines to cover all the varied conditions that may occur in an existing tank. When design and construction details are not given and are not available in the as-built standard, details that provide a level of integrity equal to the level provided by the current edition of API Standard 650 must be used.

The standard recognizes fitness-for-service assessment concepts for evaluating in-service degradation of pressure-containing components. API Recommended Practice 579, Recommended Practice for Fitness-for-Service, provides detailed assessment procedures or acceptance criteria for specific types of degradation referenced in this standard. When this standard does not provide specific evaluation procedures or acceptance criteria for a specific type of degradation or when this standard explicitly allows the use of fitness-for-service criteria, RP 579 may be used to evaluate the various types of degradation or test requirements addressed in this standard.

Table of Contents

6.10. Non-Destructive Examinations.

7. Materials.

    7.1. General.

    7.2. New Materials.

    7.3. Original Materials for Reconstructed Tanks.

    7.4. Welding Consumables.

8. Design Considerations for Reconstructed Tanks.

    8.1. General.

    8.2. New Weld Joints.

    8.3. Existing Weld Joints.

    8.4. Shell Design.

    8.5. Shell Penetrations.

    8.6. Windgirders and Shell Stability.

    8.7. Roofs.

    8.8. Seismic Design.

9. Tank Repair and Alteration.

    9.1. General.

    9.2. Removal and Replacement of Shell Plate Material.

    9.3. Shell Repairs Using Lap-Welded Patch Plates.

    9.4. Repair of Defects in Shell Plate Material.

    9.5. Alteration of Tank Shells to Change Shell Height.

    9.6. Repair of Defective Welds.

    9.7. Repair of Shell Penetrations.

    9.8. Addition or Replacement of Shell Penetrations.

    9.9. Alteration of Existing Shell Penetrations.

    9.10. Repair of Tank Bottoms.

    9.11. Repair of Fixed Roofs.

    9.12. Floating Roofs.

    9.13. Repair or Replacement of Floating Roof Perimeter Seals.

    9.14. Hot Taps.

10. Dismantling and Reconstruction.

    10.1. General.

    10.2. Cleaning and Gas Freeing.

    10.3. Dismantling Methods.

    10.4. Reconstruction.

    10.5. Dimensional Tolerances.

11. Welding.

   11.1. Welding Qualifications 1.

   11.2. Identification and Records.

12. Examination and Testing.

   12.1. Nondestructive Examinations.

   12.2. Radiographs.

   12.3. Hydrostatic Testing.

   12.4. Leak Tests.

   12.5. Measured Settlement during Hydrostatic Testing.

13. Marking and Recordkeeping.

   13.1. Nameplates.

   13.2. Recordkeeping.

   13.3. Certification.

Appendix A. Background on Past Editions of API Welded Storage Tank Standards.

Appendix B. Evaluation of Tank Bottom Settlement.

Appendix C. Checklists for Tank Inspection.

Appendix D. Authorized Inspector Certification.

Appendix E. Technical Inquiries.

Appendix F. NDE Requirements Summary.

Appendix G. Qualification of Tank Bottom Examination Procedures and Personnel.

Appendix S. Austenitic Stainless Steel Storage Tanks.

6.4   Piping—Specific Guidelines to Layout

6.4.1   Tank Grouping

A tank area is subdivided into various groups, determined by the contents of the tanks and the relative shape and area of the plot available; access and firefighting must also be considered. See Table 6–1 and Figure 6–1.

The tank grouping and layout depends on the hazard classification of the stored liquid. Table 6–2 lists the classifications of crude oil and its derivatives. Crude oil and its derivatives are potentially hazardous materials. The degree of the hazard is determined essentially by volatility and flash point.

The Institute of Petroleum has specified these classes. Refer to latest edition of *IP Refinery Safety Code Part 3*.

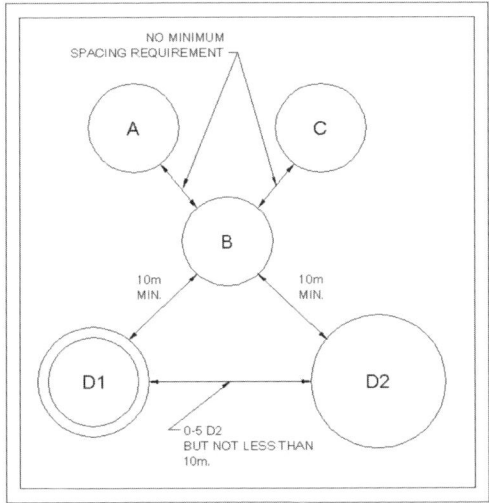

Figure 6–1 *Tanks A, B, and C are fixed or floating roof small tanks (less than 10 m diameter) with a total capacity of less than 8000 m³; no intertank spacing requirements other than for construction, operation, and maintenance convenience. Tanks D1 and D2 have diameters greater than 10 m, with the diameter of D2 greater than D1. Intertank spacing between smaller and larger tanks. The photos are of a tank farm and a spherical LPG storage tank. (Courtesy of Red Bag/Bentley Systems, Inc.)*

6.4.2 Tank Area Layout

General

The layout of tanks, as distinct from their spacing, always should take into consideration the accessibility needed for firefighting activities and the potential value of a storage tank farm in providing a buffer area between process plant and public roads, houses, and the like, for

Table 6–1   API Tank Size, for Layout Purpose

Approximate Capacity[a]		Diameter (m)	Height (m)
U.S. Barrels	Meters³		
500	75	4.6	4.9
1.000	150	6.4	4.9
1.500	225	6.4	7.3
2.000	300	7.6	7.3
3.000	450	9.2	7.3
4.000	600	9.2	9.3
5.000	750	9.2	12.2
6.000	900	9.2	14.6
7.000	1050	12.2	9.9
9.000	1350	12.2	12.2
10.000	1500	12.8	12.2
12.000	1800	12.8	14.6
15.000	2250	14.6	14.6
20.000	3000	18.3	12.2
30.000	4500	22.3	12.2
40.000	6000	26.0	12.2
50.000	7500	27.5	14.6
90.000	12000	36.6	12.2
100.000	15000	41.0	12.2
120.000	18000	41.0	14.6
140.000	21000	49.8	12.2
180.000	27000	54.9	12.2
200.000	30000	54.9	14.6
300.000	45000	61.0	17.0

Table 6–1 API Tank Size, for Layout Purpose (cont'd)

Approximate Capacity[a]		Diameter (m)	Height (m)
U.S. Barrels	Meters³		
450.000	60000	73.2	17.0
600.000	90000	91.5	14.6
800.000	100000	105.0	14.6

a. Source: Based on API 650.

Table 6–2 Institute of Petroleum Classifications of Crude Oil and Its Derivatives

Class	Sub-class	Description
Class 0		Liquified petroleum gases (LPG)
Class I		Liquids with flash points below 21°C
Class II	1	Liquids with flash points from 21°C up to and including 55°C, handled below flash point
	2	Liquids with flash points from 21°C up to and including 55°C, handled at or above flash point
Class III	1	Liquids with flash points above 55°C up to and including 100°C, handled below flash point
	2	Liquids with flash points above 55°C up to and including 100°C, handled above flash point
Unclassified		Liquids with flash points above 100°C

environmental reasons. The location of the tank area relative to process units must be such as to ensure maximum safety from possible incidents.

Primarily, the requirements for the layout of refinery tanks farms are summarized as follows:

1. Intertank spacing and separation distances between tanks and the boundary line and other facilities are of fundamental importance.

2. Suitable roadways should be provided for approach to tank sites by mobile firefighting equipment and personnel.

3. The fire water system should be laid out to provide adequate fire protection to all parts of the storage area and the transfer facilities.

4. Bunding and draining of the area surrounding the tanks should be such that spillage from any tank can be contained and controlled to minimize subsequent damage to the tank and its contents. They also should minimize the possibility of other tanks being involved.

5. Tank farms preferably should not be located at higher levels than process units in the same catchment area.

6. Storage tanks holding flammable liquids should be installed in such a way that any spill will not flow toward a process area or any other source of ignition.

Spacing of Tanks for LPG Stocks of Class 0

The distance given in Table 6–3 are the minimums recommended for aboveground tanks and refer to the horizontal distance in plan between the nearest point on the storage tank and a specified feature, such as an adjacent storage tank, building, or boundary. The distances are for both spherical and cylindrical tanks. Refer to latest edition of *IP Refinery Safety Code Part 3*.

Bunding and Grouping of LPG Tanks

The provision of bunds around LPG pressure storage tanks normally is not justified. Separation curbs should be low to avoid gas traps, maximum 600 mm high; and they may be located to prevent spillage reaching important areas, such as the pump manifold area or pipe track. The ground under the tanks should be graded to levels that ensure any spillage has a preferential flow away from the tank. Pits and depressions, other than those provided as intentional catchment areas, should be avoided to prevent the forming of gas pockets.

Pressure storage tanks for LPG should not be located within the bunded enclosures of Class I, II, or III product tankage or low-pressure refrigerated LPG tankage. The layout and grouping of tanks, as distinct from spacing, should receive careful consideration with the view of accessibility for firefighting and the avoidance of spillage from one tank flowing toward another tank or a nearby important area.

Table 6–3   Spacing of Tanks for LPG Stocks of Class 0

Factor	Recommended Spacing for LPG Stored in Pressure Tanks
1. Between LPG pressure storage tanks	One quarter of the sum of the diameters of the two adjacent tanks
2. To Class I, II, or III product tanks	15 m from the top of the surrounding Class I, II or III product tanks
3. To low-pressure refrigerated LPG tanks	One diameter of the largest low pressure refrigerated storage tanks but not less than 30 m
4. To a building containing flammable material, such as a filling shed, storage building	15 m
5. To a boundary or any fixed source of ignition	Related to the water capacity of tank as follows:

Capacity:	Distance:
Up to 135 m^3	15 m
Over 135 to 565 m^3	24 m
Over 565 m^3	30 m

Spacing of Tanks for Low-Pressure Refrigerated LPG Storage Class 0

The distances given in Table 6–4 are the minimum recommended and refer to the horizontal distance in plan between the nearest point on the storage tank and a specified feature, such as an adjacent storage tank or building, boundary. Refer to latest edition of *IP Refinery Safety Code Part 3*.

Bund or Impounded Basin for Refrigerated LPG Storage

A bund should be provided around all low pressure tanks containing refrigerated LPG. The tank should be completely surrounded by the bund, unless the topography of the area is such, either naturally or by construction, that spills can be directed quickly and safely, by gravity drainage and diversion walls if required, to a depression or impounding basin located within the boundary of the plant.

Bunds should be designed to be of sufficient strength to withstand the pressure to which they would be subjected if the volume within the bunded enclosure were filled with water, either from firefighting equipment or rainwater. The area within the bund, depression, or

Table 6–4   Spacing of Tanks for Low-Pressure Refrigerated LPG Storage Class 0

Factor	Recommended Spacing for Low Pressure Refrigerated LPG Storage
1. Between refrigerated LPG storage tanks	One half the sum of the diameters of the two adjacent tanks
2. To Class I, II, or III product tanks	One diameter of the largest refrigerated storage tank but no less than 30 m
3. To pressure storage tanks	One diameter of the largest refrigerated storage tank but no less than 30 m
4. To process units, office building, workshop, laboratory, warehouse, boundary, or any fixed source of ignition	45 m

impounding basis should be isolated from any outside drainage system by a valve, normally closed unless the area is being drained of water under controlled conditions.

Where only one tank is within the bund, the capacity of the bunded enclosure, including the capacity of any depression or impounding basis, should be 75% of the tank capacity. Where more than one tank is within the main enclosure, intermediate bunds should be provided, so as to give an enclosure around each tank of 50% of the capacity of that tank, and the minimum effective capacity of the main enclosure, including any depression or impounding basin, should be 100% of the capacity of the largest tank, after allowing for the volume of the enclosure occupied by the remaining tanks. It is desirable for the required capacity to be provided with bunds not exceeding an average height of 6 ft as measured from the outside ground level. The area within the bund should be graded to levels that ensure any spillage has a preferential flow away from the tank.

No tankage other than low-pressure tankage for refrigerated LPG should be within the bund. The layout and grouping of tanks, as distinct from spacing, should receive careful consideration with the view of accessibility for firefighting.

Spacing of Tanks for Petroleum Stocks of Classes I, II, and III(2)

Table 6–5 gives a guidance on the minimum tank spacing for Class I, II, and III(2) storage facilities. Refer to latest edition of *IP Refinery Safety Code Part 3*.

The following points should be noted:

1. Tanks of diameter up to 10 m are classed as small tanks.
2. Small tanks may be sited together in groups, no group having an aggregate capacity of more than 8000 m^3. Such a group may be regarded as one tank.
3. Where future changes of service of a storage tank are anticipated, the layout and spacing should be designed for the most stringent case.
4. For reasons of firefighting access, there should be no more than two rows of tanks between adjacent access roads.
5. Fixed roof with internal floating covers should be treated for spacing purposes as fixed roof tanks.
6. Where fixed roof and floating roof tanks are adjacent, spacing should be on the basis of the tank(s) with the most stringent conditions.
7. Where tanks are erected on compressible soils, the distance between adjacent tanks should be sufficient to avoid excessive distortion. This can be caused by additional settlement of the ground where the stressed soil zone of one tank overlaps that of the adjacent tank.
8. For Class III(1) and unclassified petroleum stocks, spacing of tanks is governed only by constructional and operational convenience. However, the spacing of Class III(1) tankage from Class I, II, or III(2) tankage is governed by the requirements for the latter.
9. For typical tank installation, illustrating how the spacing guides are interpreted, see Figures 6–2 through 6–4.

For the details of a typical vertical tank foundation, see Figures 6–5 and 6–6.

Compound Capacities

Aboveground tanks for Class I, II(1), II(2), and III(2) petroleum liquids should be completely surrounded by a wall or walls. Alternatively, it is acceptable to arrange that spillage or a major leak from any tank be directed quickly and safely by gravity to a depression or impounding basin at a convenient location.

Table 6–5   Spacing of Tanks for Petroleum Stocks of Classes I, II, and III(2)

Factor	Type of Tank Roof	Recommended Minimum Distance
1. Within a group of small tanks	Fixed or floating	Determined solely by construction or maintenance operational convenience
2. Between a group of small tanks or other larger tanks	Fixed or floating	10 m minimum, otherwise determined by the size of the larger tanks (see factor 3)
3. Between adjacent individual tanks (other than small tanks)	Fixed	Half the diameter of the larger tank but no less than 10 m and need not be more than 15 m
	Fixed	0.3 times the diameter of the larger tank but no less than 10 m and need not be more than 15 m (in the case of crude oil tankage, the 15 m option does not apply)
4. Between a tank and the top of the inside of the wall of its compound	Fixed or floating	Distance equal to no less than half the height of the tank (access around the tank at compound grade level must be maintained)
5. Between any tank in a group of tanks and the inside top of the adjacent compound wall	Fixed or floating	
6. Between a tank and a public boundary fence	Fixed or floating	No less than 30 m
7. Between the top of the inside of the wall of a tank compound and a public boundary fence or to any fixed ignition source		No less than 15 m
8. Between a tank and the battery limit of a process plant	Fixed or floating	No less than 30 m
9. Between the top of the inside wall of a tank compound and the battery limit of a process plant		No less than 15 m

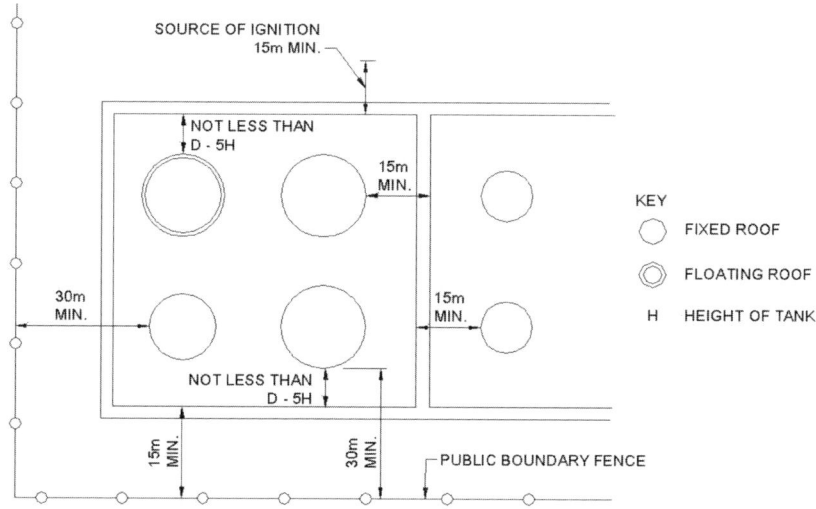

Figure 6–2 *Tank and compound wall distances from typical features (courtesy of Red Bag/Bentley Systems, Inc.).*

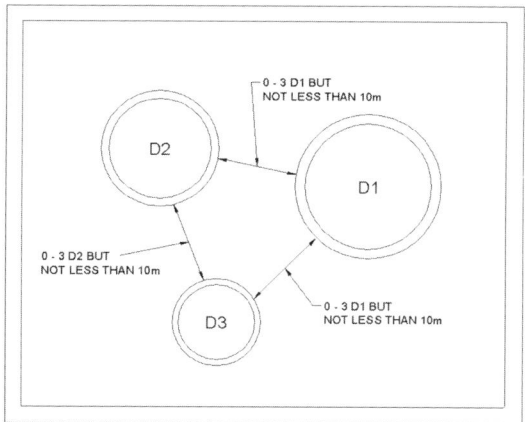

Figure 6–3 *Floating roof tanks of diameter D1, D2, and D3 are greater than 10 m within the same compound; D1 is greater than D2, and D2 is greater than D3. (Courtesy of Red Bag/Bentley Systems, Inc.)*

The distance between the edge of the impounding basin and the nearest tank or the inside top of the nearest bund wall should be a minimum of 30 m. The distance between the edge of the basin and road fence battery limit of a process plant should be no less than 15 m.

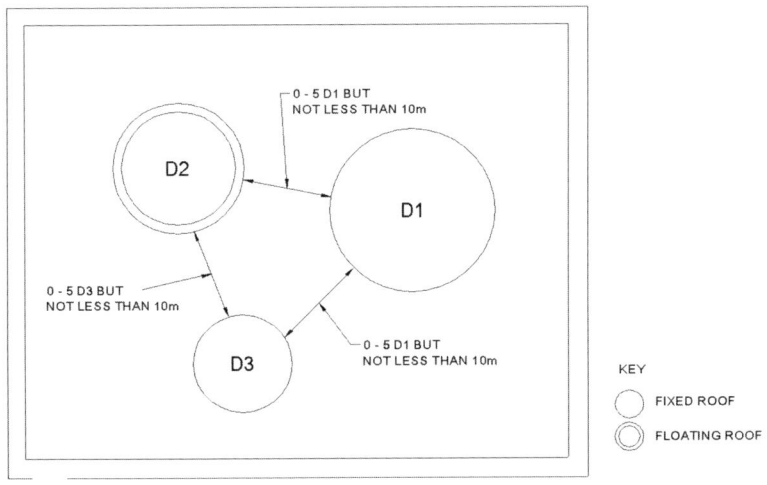

Figure 6–4 *Intertank spacing for floating roof tanks (greater than 10 m diameter). Fixed and floating roof tanks are within the same compound; D1 is greater than D2, D2 is equal to D3. (Courtesy of Red Bag/Bentley Systems, Inc.)*

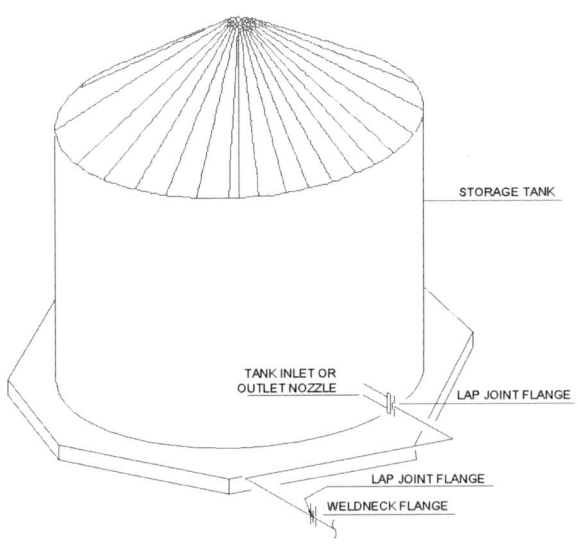

Figure 6–5 *Lap joint flange detail for tank settlement (courtesy of Red Bag/Bentley Systems, Inc.).*

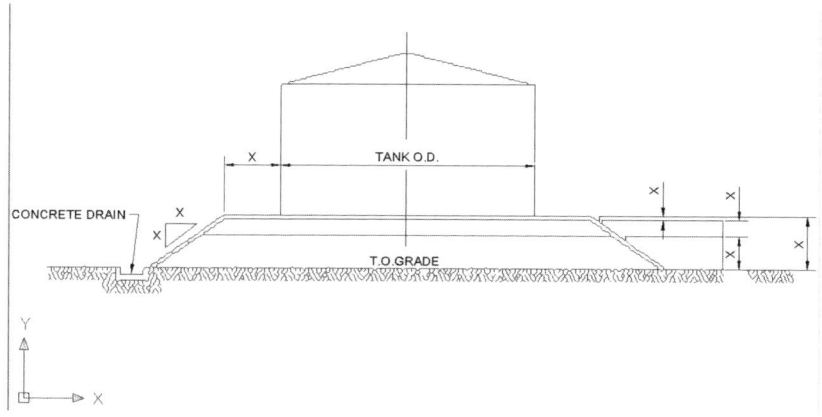

Figure 6–6   *Foundation for vertical tank, based on BS2654 (courtesy of Red Bag/Bentley Systems, Inc.).*

The height of the bund wall as measured from outside ground level should be sufficient to afford protection for personnel when engaged in firefighting, and the wall should be located so that a reasonably close approach can be made to a tank fire to allow use of mobile fire-fighting equipment. Access roads over bund walls into very large compounds are helpful in certain fire situations.

Separate walls around each tank are not necessary, but the total capacity of the tanks in one bunded area should be restricted to the maximum figures in Table 6–6. Refer to latest edition of *IP Refinery Safety Code Part 3*. The figures for the latter two may be exceeded for groups of no more than three tanks, where there can be no risk of pollution or hazard to the public.

Intermediate walls of lesser height than the main bund walls may be provided to divide tank areas into groups of a convenient size so as to contain small spillages and act as fire breaks.

Table 6–6   Tank Capacity in a Single Bunded Area

Single Tanks	No Restriction
Groups of floating roof tanks	120,000 m^3
Groups of fixed roof tanks	60,000 m^3
Crude tanks	No more than two tanks of greater individual capacity than 60,000 m^3

Buried, semi-buried, or mounded tanks need not be enclosed by a bund wall except when they are located in ground higher than the surrounding terrain. However, consideration should be given to the provision of small bund walls around associated tank valves.

The net capacity of the tank compound generally should be equivalent to the capacity of the largest tank in the compound. However, a reduction of this capacity to 75% provides reasonable protection against spillage and may be adopted where conditions are suitable (e.g., where there can be no risk of pollution or hazard to the public). The net capacity of a tank compound should be calculated by deducting from the total capacity the volume of all tanks, other than the largest, below the level of the top the compound wall and the volume of all intermediate walls.

A low wall, which need be no more than 0.5 m high, should be constructed for Class III 1) and unclassified petroleum product tank areas where conditions are such than any spillage or leakage could escape from the installation and cause damage to third-party property, drainage systems, rivers, or waterways. Where there is the possibility that tanks storing these products may be required in the future for Class I, II(1), or III(2) products, then the compound walls should be suitable for this potential situation.

Pump Areas

Pumps are located outside bund areas. The practice is to group the pumps into bays, keeping the suction lines as short as practical. The discharge piping runs on low level tracks to the process or loading areas. These tracks usually pass under roads in culverts but may pass over them on a pipe bridge. Long pipe runs may require expansion loops to provide flexibility.

Fire Protection

For storage areas, the major firefighting effort is provided by mobile equipment laying down large blankets of foam or applying large volumes of water for cooling purposes. It is essential to provide a good system of all-weather roads to facilitate the transfer of fire protection materials and equipment to the scene of the fire. These roads must be of adequate width and, wherever possible, with no dead ends.

It is important in siting tanks, bund walls, and access roads, that the tanks can be protected by cooling water or foam appliances situated outside the compound walls. Account must be taken of the height of the tank and the possible need to cool the roof or project foam onto a tank. Dry risers for foam may be provided to the top of storage tanks with their connections adjacent to access roads, fixed monitors may also be employed. The flow diagram will define the system to be employed.

Road and Rail Loading Facilities

Road and rail loading facilities usually associated with storage areas. The safe location of these in relation to storage tanks is laid down in previous sections. The road or rail car is filled from a loading island, the supply lines are either routed underground or on an overhead pipe bridge. Check for clearances.

It has become common practice to provide a vapor collection system for the safe removal of vapor during the loading process. This system employs unloading arms connected to a collection system and piped to a vent stack at a safe location.

When laying out a loading area, consideration must be given to the number of vehicles or rail cars to be loaded per hour. A suitable movement pattern must be established for incoming and outgoing vehicles or railcars. Weigh bridges are required; the system of moving rail cars must be determined; building housing, operation offices, facilities for drives, and the like must be provided.

6.4.3   Tank Farm Piping and Layout

- Pipelines connected to tanks should be designed so that the stresses imposed are within the tank design limits.

- The settlement of the tank and the outward movement of the shell under the full hydrostatic pressure should be taken into account.

- The first pipe support from the tank should be located at a sufficient distance to prevent damage to both the line and tank connections.

- Consideration may be given to installing spring supports near the tank connection for large bore pipework.

- As large-diameter tanks have a tendency to settle in their foundations, provision must be made in the suction and filling piping to take care of tank settlement. This may require the use of expansion joints, victaulic couplings, or a lap joint flange installed as shown in Figure 6–5.

- The following note should be considered to all piping drawings containing storage tanks: All piping must be disconnected from the tank during hydrostatic test of the storage tank.

- The number of pipelines in tank compounds should be kept to a minimum, and they should be routed in the shortest practicable way to the main pipe tracks located outside the tank compounds, with adequate allowance for expansion.

- Flexibility in piping systems may be provided through the use of bends, loops, or offsets.

- Where space is a problem, suitable expansion joints of the bellows type may be considered for installation in accordance with manufacturer's design specifications and guidelines. These expansion joints should be used only in services where the fluid properties are such that it cannot plug the bellows. They should be anchored and guided, should not be subjected to torsional loads, and should be capable of ready inspection.

- Tank farm piping preferably should be run above ground on concrete or steel supports.

- The ground beneath piping should be so graded as to prevent the accumulation of surface water or product leakage.

- Manifolds should be located outside the tank bunds.

- Piping should pass over earth bund walls; however, if this is impossible and the pipe passes through the bund, then a suitable pipe sleeve should be provided to allow for expansion and possible movement of the lines. The annular space should be properly sealed. Likewise, lines passing through concrete bund walls should be provided with pipe sleeves.

- Pedestrian walkways should be provided to give operational access over ground-level pipelines.

- Pipelines should be protected against uneven ground settlement where they pass under roadways, railways, or other points subject to moving loads.

- Buried pipelines should be protected externally by corrosion-preventing materials or cathodic means.

- Routes of buried pipelines should be adequately marked and identified aboveground and recorded.

- Pipe racks carried across paths or roads should have adequate clearance from grade to allow the passage of vehicles and construction equipment.

- Adequate access stairways or ladders and operating platforms should be provided to facilitate operation and maintenance at tanks.

- Tanks may be interconnected at roof level by bridge platforms.

- All nozzles, including drains on a tank shell, should have block valves adjacent to the tank shell or as close as practicable.
- Precaution must be taken to prevent drain or sample valves freezing in the open position.
- The flow diagram indicates the type of double valving to be installed, with a minimum distance between the valves of 1 m. Do not allow liquid traps in vent lines.

6.4.4 Piping Support and Stress Issues

Liquid and vapor pipelines should have adequate flexibility to accommodate any settlement of tanks or other equipment, thermal expansion, or other stresses that may occur in the pipework system. The nozzle load should not exceed the maximum loads as stated by the tank manufacturer.

CHAPTER 7

Columns

7.1   Introduction

The column or tower is a vertical vessel with associated internals. The column's function usually is to distill fractions of raw product, such as crude oil, and uses the difference in boiling temperatures of the various fractions in a raw product. The vessel internals can be trays or a packing. The design considerations described in this chapter is intended for columns with trays; however, it also is applicable to piping design associated with packed columns.

The column process system generally is composed of accompanying equipment such as reboilers and condensers, and the column should not to be regarded as an isolated piece of equipment for the purposes of piping design. Therefore, it is necessary to design the column with sufficient consideration given to the relationships among the other items of equipment

7.2   Internals

Column internals are used to direct the internal flow of the fluids and vapors. The internals enhance the separation or fractionation process. They improve the contact between the vapors and the liquids. Refer to Figure 7–3A for the most common terms used with columns: overhead, reflux, feed, return, and bottoms.

The fractionation column constitutes a vertical cylinder interspaced at equal intervals with trays. Often the trays are in the form of simple perforated discs, but more complex arrangements occur, depending on column function. Contents are heated near the bottom, producing vapors that rise through the trays, meeting a flow of liquid passing down as a result of condensation occurring at various levels.

It is essential to ensure maximum surface contact between the vapor and liquid. The lightest fractions are drawn from the highest elevations or overheads, the heaviest residue being deposited at the bottom.

The overheads are cooled down by an overhead condenser and the liquids are returned as reflux at the top of the column to extract possible remaining fractions of the overhead. The liquids travel through the column internals downward.

The bottoms are heated by the bottoms reboiler and the vapors are returned to the column at the reboiler return nozzle, to pass through the column upward.

The feed of raw material is in the middle sections of the column, where the liquids travel downward through the column internals. By contacting the heated vapors coming from the bottom, the first volatile vapors start to separate from the feed.

7.3   Applicable International Code

Numerous international codes and standards that apply to the various types of columns and vessels used in hydrocarbon processing plants and listed here are several of the most important, along with their scope and table of contents. The design and specifying of these items of equipment are the responsibility of the mechanical engineer; however, a piping engineer or designer will benefit from being aware of these documents and reviewing the appropriate sections that relate directly to piping or a mechanical-piping interface.

7.3.1   API Specification 12J. Specification for Oil and Gas Separators

Scope

This specification covers the minimum requirements for the design, fabrication, and shop testing of oilfield-type oil and gas separators and oil-gas-water separators used in the production of oil or gas, usually located but not limited to some point on the producing flowline between the wellhead and pipeline. Separators covered by this specification may be vertical, spherical, or single- or double-barreled horizontal.

Unless otherwise agreed on between the purchaser and the manufacturer, the jurisdiction of this specification terminates with the pressure vessel as defined in the scope of Section VIII, Division 1 of the *ASME Boiler and Pressure Vessel Code,* hereinafter referred to as the

ASME Code. Pressure vessels covered by this specification normally are classified as natural resource vessels by API 510, *Pressure Vessel Inspection Code.* Separators outside the scope of this specification include centrifugal separators, filter separators, and desanding separators.

Table of Contents

Foreword.

Policy.

Section 1. Scope.

Section 2. Definitions.

Section 3. Material.

Section 4. Design.

Section 5. Fabrication, Testing, and Painting.

Section 6. Marking.

Section 7. Inspection and Rejection.

Appendix A. Process Considerations.

Appendix B. Corrosion Guidelines.

Appendix C. Design and Sizing Calculations.

Appendix D. Separator Sizing Example Calculation.

Appendix E. Separator Design Information.

Appendix F. Use of Monogram.

7.4 Piping—Specific Guidelines to Layout

7.4.1 General

- It is necessary to establish whether a column is working as a single unit or in conjunction with similar others. Depending the process arrangements, flow may be sequential from one to the next. Therefore, for maximum economy, the order of columns must be arranged to give shortest piping runs and lowest pumping losses.

- Consideration should be given, where necessary, to the material used, since although correct sequence may have been affected, unnecessary expense may be involved with alloy lines. Such cases must be treated on their merits.

- The column usually is interconnected with other process equipment; therefore, it should be located adjacent to the pipe rack to allow piping connections to run to and from the rack in an orderly fashion.

- Automatically this means that walkways (provided for the erection of trays and maintenance) should be located on the side of the column away from the rack to provide suitable access for equipment required to be removed (see Figures 7–1 and 7–2). This is not mandatory, since occasions arise when other arrangements are necessary.

7.4.2   Internals

- Having located the walkways, orientate the internal trays to ensure unimpeded access.
- Depending on the type of tray used, one or more downcomer partitions may be required.
- If these are orientated incorrectly entry will be impossible via manhole unless the column is exceptionally large.
- Preferably downcomers are arranged normal to the walkway access center line.
- Process connections can be located in a logical sequence, starting from the top, when the tray orientation is established.

7.4.3   Overheads

- Highest connection is the overhead vapor line, which usually is connected to a condensing unit. An air fin unit may be located on top of the pipe rack.
- Alternatively, a shell and tube exchanger condenser usually is located on a structure adjacent to the column (as may an air fin unit).
- The overhead vapor connection may project vertically from the top of the column followed by a 90° bend or it may emerge from the side at 45°, using an integral projection extending up almost to the center top inside the head (see Figure 7–3).
- The latter is more economic in piping, since it eliminates the use of some expensive fittings, especially when large-diameter overhead lines are involved. It also makes piping supporting simpler, as use of a 45° elbow enables the pipe to run down close to the column. A disadvantage is greater rigidity making stress analysis more difficult.

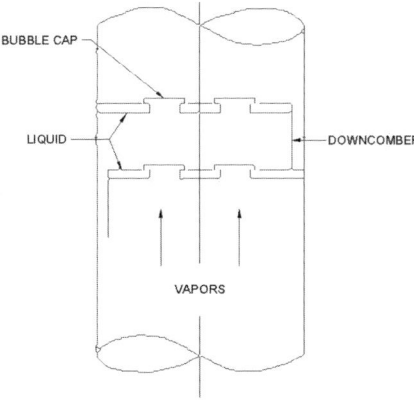

Figure 7–1   *Section (vertical) through column (courtesy of Red Bag/Bentley Systems, Inc.).*

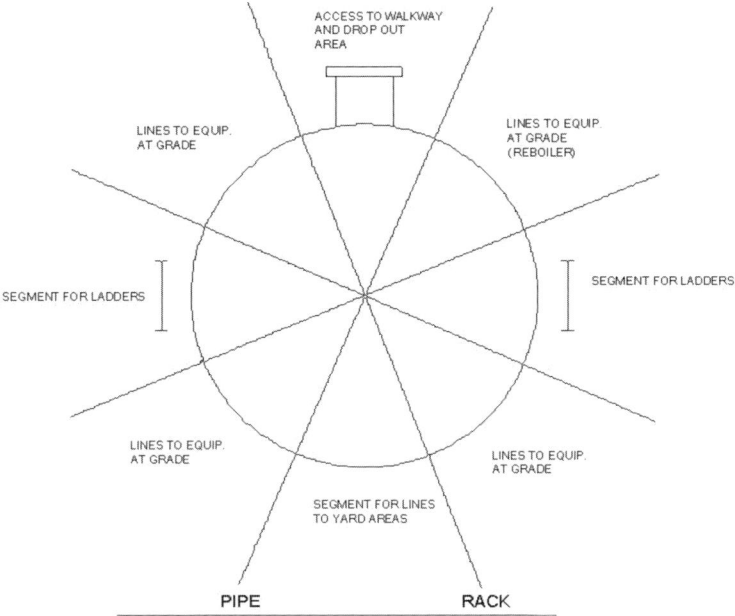

Figure 7–2   *Section (horizontal) through column (courtesy of Red Bag/Bentley Systems, Inc.).*

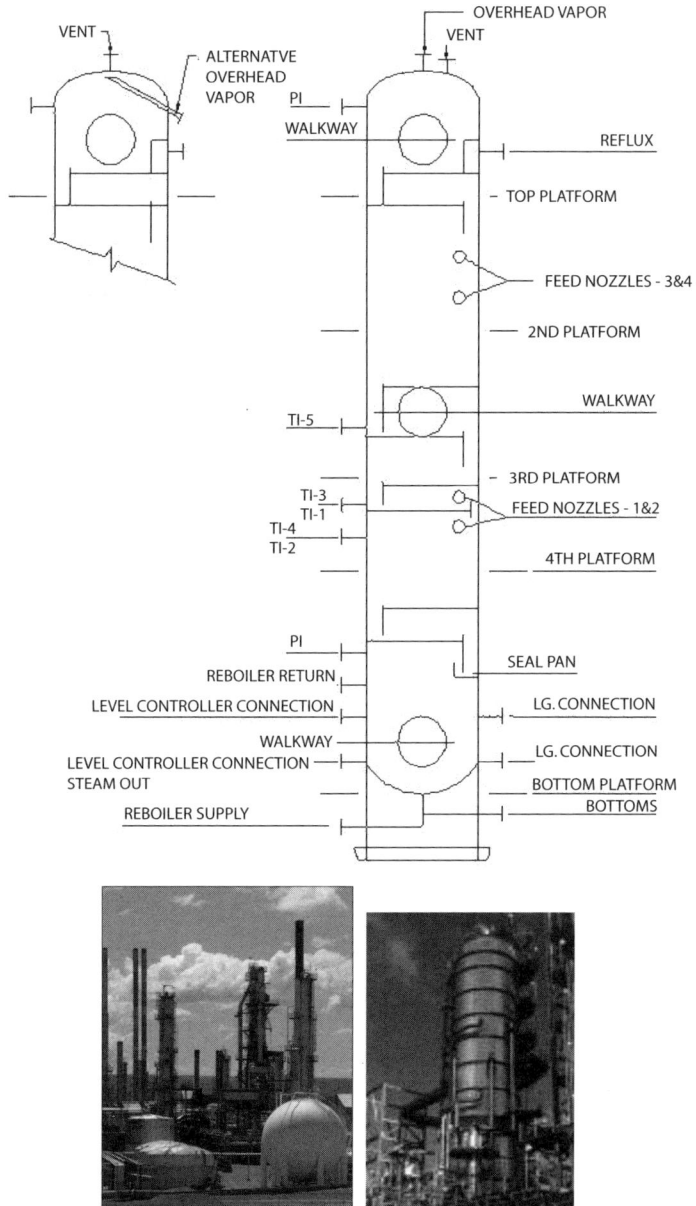

Figure 7–3   *View of column. The photos show tanks and spheres with distillation columns in the background and an FCC distillation column. (Courtesy of Red Bag/Bentley Systems, Inc.)*

- Often, the permitted pressure drop between outlet nozzle and exchanger is low (i.e., approximately 0.5 psi); hence, it is essential to obtain the straightest and shortest run possible. If the connection is from the top, special arrangements must be made to support the line, which often is of large diameter.
- The condenser usually is self-draining.
- Often, some of the condensed liquid must be pumped back into the column (reflux). The condensed liquid flows by gravity to a reflux drum situated immediately below the exchanger.
- This drum also can be mounted in the structure supporting the exchanger. This is important, since if it were located elsewhere, an additional pump would be required if the liquid had to be elevated again after flowing from the condenser.
- Furthermore, since the liquid in the reflux drum has to be returned to the column by a pump, it is convenient to locate this underneath the reflux drum at the base of the structure.
- It follows, therefore, that the orientation of the outlet of the vapor connection automatically fix the location of the exchanger and the reflux drum or vice versa.
- The reflux pump discharges back into the tower, usually at a high elevation; and since the piping should be as straight and short as possible, it will probably come up at the side of the vapor connection.

7.4.4 Reflux

- Trays are numbered starting from the top.
- The first downcomer therefore is an odd one.
- Often, the reflux connection is located above the top tray (see Figure 7–4). This means that orientation of the odd and even trays can be fixed, since to utilize the tray contact surface, the reflux connection must occur on the opposite side of the downcomer.

7.4.5 Feeds

- The most important connections are the feeds (see Figures 7–4 through 7–8). These may occur over the odd or even trays or a combination of both.
- Their elevation location is entirely a function of the process design. The orientation of the nozzle follows the elevation location, but ensure that the nozzle is not located behind the

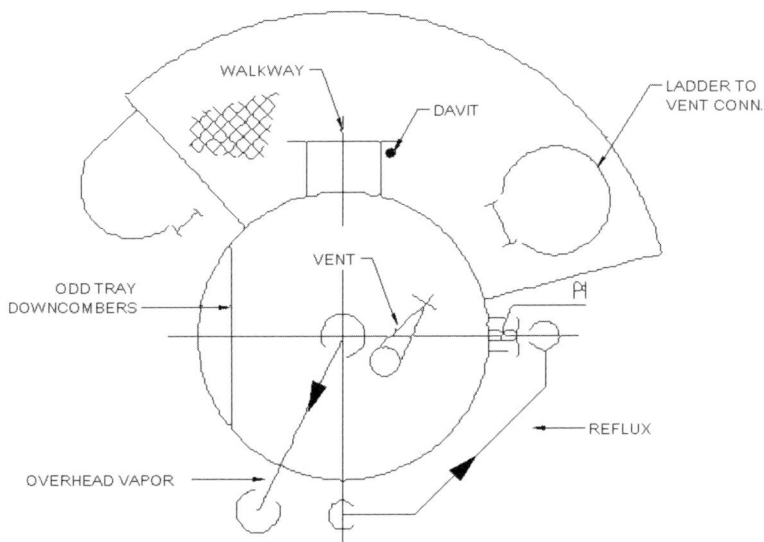

Figure 7–4   *Top platform (courtesy of Red Bag/Bentley Systems, Inc.).*

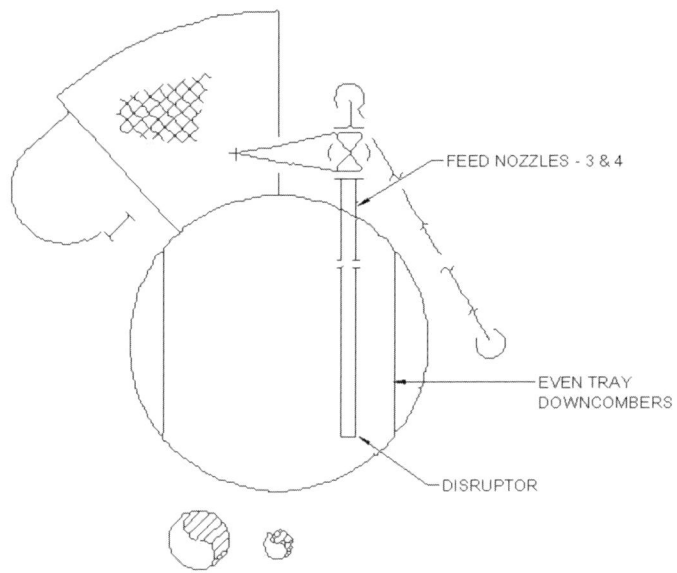

Figure 7–5   *Second platform (courtesy of Red Bag/Bentley Systems, Inc.).*

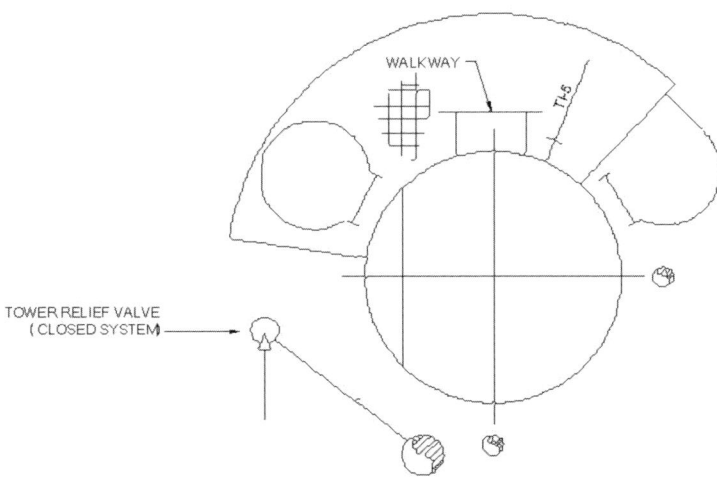

Figure 7–6   *Third platform (courtesy of Red Bag/Bentley Systems, Inc.).*

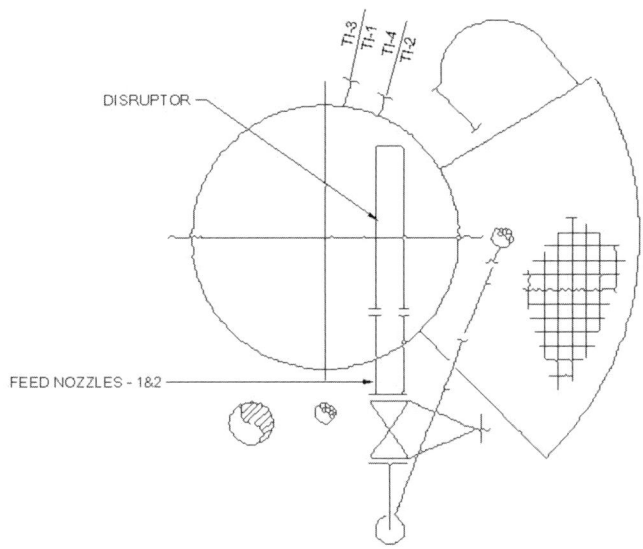

Figure 7–7   *Fourth platform (courtesy of Red Bag/Bentley Systems, Inc.).*

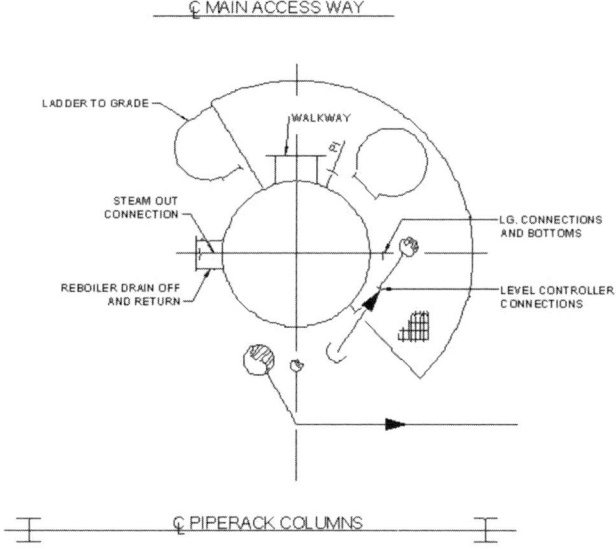

Figure 7–8   *Bottom platform (courtesy of Red Bag/Bentley Systems, Inc.).*

downcomer from the tray above. This would result in a liquid buildup in the downcomer, restricting flow through the tray below and preventing correct operation.

- Feed connections usually are located in the tray area between the downcomers in a manner to ensure maximum use of the tray area as possible. Often an internal feed pipe or sparger is used to facilitate this.

7.4.6   Instruments

- Instrument connections must be arranged so that pressure connections are in the vapor space and temperature connections are immersed in the liquid.
- Sometimes, it is better to put the temperature connections in the downcomer part of the tray, since the depth of the liquid is greater and this practice easily obtains effective coverage.
- The probe length and mechanical interference may prevent this; if so, locate thermowell over the tray itself. This should be done only on installations where the liquid depth on the tray is sufficient.
- The base of the tower contains a level of liquid controlled by high- and low-level controllers, liquid level alarms, and level gauges (see Figure 7–8).

- Care should be taken when orientating these instruments that they do not obstruct access on the platform.

- The physical space that these instruments occupy can be excessive. Do not position immediately adjacent to ladders or manholes.

7.4.7 Reboiler Connections

Reboiler liquid and vapor connections are located relevant to either the liquid head (elevation) or stress requirements or both. Two configurations are possible, vertical and horizontal.

- For horizontal reboilers, consideration must be given to the shortest, most direct connection route to reduce pressure drop, which probably will govern design layout. In both cases, there may be support problems. Usually, a vertical reboiler, thermosyphon operated, offers the easiest solution.

- Flexibility is obtained on the lower connection, where entry from the bottom of the tower can be varied as required and support problems are simplified.

7.4.8 Platforms

- All the previous requires access of some kind. To minimize cost, a minimum amount of platforms should be provided consistent with safety.

- Orientation arrangements should be designed to avoid need for 360° platforms.

- A platform should not extend almost entirely round the column simply to provide access to a temperature connection that could have been located on the opposite side.

- Where several columns may work together, link platforms may be required to move from one to the next. In these cases, strict consideration must be given to maximize economy and the differential growth of each column.

- Overall height is governed by the number of trays, the pump net positive suction head requirements, and the static liquid head. This last head, necessary for the thermosyphon reboiler, establishes the skirt height.

- Some circumstances require routing lines partially around the column. Should these lines cross a platform, sufficient headroom clearance must be provided.

7.4.9   Relief Valves

- Relief valves vary in number and size. Location can depend on whether the discharge is to the atmosphere or a closed system.
- If discharging to a closed system, locate the relief valves at a convenient platform down the column, above the relief header.
- If discharging to the atmosphere, locate the relief valves on top of the column, with the open end of the discharge a minimum of 3000 mm above the top platform.
- For maintenance removal, the valve should be located to allow a top head davit to pick it up. Depending on size, multiple relief valves may require a gantry structure on the top head.

7.4.10 Clips

- Early orientation of the nozzles and routing of the lines allow platforms and pipe support clip locations to be passed quickly to the column vendor.
- It is becoming more a requirement that clips be welded on in the vessel fabrication shop, particularly for alloy steels and heat-treated vessels.
- When locating platforms and ladders, the maximum ladder length should not exceed 9 m without a rest platform.
- Platforms should be elevated 900 mm below walkways, where possible.
- Walkway davits or hinges should be positioned to avoid the swing of a cover fouling instruments or other connections.
- When positioning vertical piping, to simplify supporting, attain a common back of pipe dimension from the column shell of 300 mm.

CHAPTER 8

Cooling Towers

8.1   Introduction

Cooling water is an essential service in any chemical plant or refinery, and control of the temperature plays a critical part in any plant process. Therefore, any water used for cooling picks up heat from the medium being cooled and must itself be cooled before being recirculated. The cooling tower enables this water cooling to be carried out.

Regardless of type of tower selected, there always is a reservoir of water at the base of tower, from which water is drawn and pumped around the plant. It is returned, via a header pipe, back to the top of the tower.

The water then is dispersed over the whole area of the tower by means of wooden slats or sprinkler nozzles. This breaks the water up into fine droplets, similar to rain, thus exposing a greater surface area and enabling cooling to be much more effective. The cooled water is collected in the basin under the tower and is ready for reuse.

8.2   Types of Cooling Towers

The two most commonly types of towers used in refineries and chemical plants operate via natural draft (venturi or chimney type) or induced draft (box type).

8.2.1   Natural Draft Cooling Towers

The natural draft type is not often used and normally is used only when the contours of the ground provide a high position on which they can be located. The higher position gives unimpeded exposure to the cooling tower.

8.2.2 Induced Draft Cooling Towers

Induced draft cooling towers are furnished in two types, based on the direction of air flow relative to the water flowing through the tower: cross flow or counter flow. In the cross flow cooling tower, the sides are entirely open; and air is passed through the sides to a central plenum chamber, across the downward flow of water, and exhausted through the top of the structure by one or more fans. Some characteristics of cross flow towers, compared to counter flow towers, are

- The contact surface is less effective.
- The air flow quantity is greater.
- Icing is more of a problem in winter months.
- The fan horsepower may be higher.

The counter flow tower has straight enclosed sides, except for an air entrance near the bottom. Air is taken in at the bottom of the tower, raised countercurrent to the downward flow of water, and exhausted at the top by means of one or more fans. Some characteristics of counter flow towers, compared to cross flow towers, are

- Lower air flow quantity.
- The fan horsepower may be lower.
- Generally, lower fire protection cost.
- Usually, lower pumping height.
- Frequently, a larger basin area.

8.3 Inlet and Outlet Piping

8.3.1 Cooling Tower Structure

The cooling tower usually is a clad wooden structure constructed with a great number of lightweight plastic components. This makes it susceptible to fire, and a fire protection system should be considered.

This lightness of construction means that nozzle forces should be at a minimum and the flexibility of piping layout is of great importance.

Access should be adequate for the maintenance of the fan motors mounted on top of the structure and sufficient to give access to any doors or hatches in the fan stacks and floor on top of the structures.

8.3.2 Layout

A cooling tower is one of the larger items of equipment, in terms of ground area, that must be located on a site plan. Factors affecting the location of cooling towers, other than convenience to water supply and return, are the prevailing wind, noise, and access roads.

Prevailing Winds

Cooling towers should be located with their small side toward the prevailing wind. The gives both long sides an equal intake of fresh air.

Cooling towers should not be downwind or adjacent to fired heaters, flare stacks, or any heat-producing items, as these raise the ambient air temperature and reduce the towers cooling efficiency.

Noise

Noise levels of larger cooling towers can be quite high and may become objectionable if the tower is located too close to continually occupied work areas, such as offices and control buildings.

Access Roads

Access is required for the essential maintenance of pumps, chemical dosing equipment, and handling trash screens.

Cooling towers lose water by evaporation and entrainment, resulting in a water spray and fog on the downwind side of the tower, making any roads continually wet. This is a traffic hazard, because in certain locations, ice can form in winter.

In general, a minimum distance of 15–20 m of clear area should be allocated for air movement about the tower.

8.3.3 Cooling Basin Design and Piping

Water Makeup

Water makeup to a cooling tower is necessary to replace the mechanical carryout of water droplets (windage), evaporation, and the blowdown required to maintain a controlled solids buildup. Makeup water usually is added to the cooling tower basin.

Control of the water level in the cooling tower basin is via a level instrument of some description. This should be located in the relatively still waters of the pump basin. If the instrument is of the level displacer type, it should be housed in a "still well" located in the pump basin. This protects the instrument and dampens the turbulent water to give a smoothed water level for measurement.

The blowdown rate depends on the solids entering in the makeup water and the solids level to be maintained in the system. Blowdown is measured by a flow indicator at any point in the cooling water circulation system that may be convenient for its disposal to a sewer.

Trash Screen and Gate

A course filter or screen should be located between the cooling tower basin and pump basin to trap floating debris and where it can be reached for regular cleaning. A submerged orifice is a useful way of trapping floating debris. By keeping the opening a few centimeters off the tower basin floor, a mud trap is formed. This prevents any silt or submerged objects from approaching the pump suction. The submerged opening is a convenient place for locating a means of isolating the pump basin from the cooling tower basin if required. This can be a proprietary penstock, or a simple wooden sluice gate.

The Pump Basin

The pump basin is an integral part of the tower basin, being cast directly onto one of its sides. Where more than one pump draws from the same basin, the chamber should be shaped or provided with baffles to prevent one pump intake affecting the flow to the others. If any sudden changes in the flow path within the pump basin do occur, the pump intake should be located at least five pipe diameters downstream from them (see Figure 8–1).

Suction Piping

High losses at the pump intake can cause excessive turbulence that may adversely affect the pump's performance. A "bell mouth" is the most effective way of reducing these losses and can simply be a concentric reducer, the large end being 1.5 times the diameter of the smaller suction pipe diameter. Clearance between the face of the "bell mouth" and the pump basin floor should be equal to the larger diameter of the "bell mouth."

Another useful aid in reducing turbulence within in the suction piping is to have at least three pipe diameters of straight pipe upstream of the pump suction inlet flange. The submerged depth of the intake usually is not very critical, but a minimum of 1 m is good practice. For vertical immersed pumps, use the vendor's recommendations. The suction line should rise positively to the pump flange to prevent air pockets. For double suction pumps, bends in the horizontal should be greater than three pipe diameters upstream of the pump flange.

If the pump is mounted on its own plinth, a check should be made with the Civil Department for possibility of differential settle-

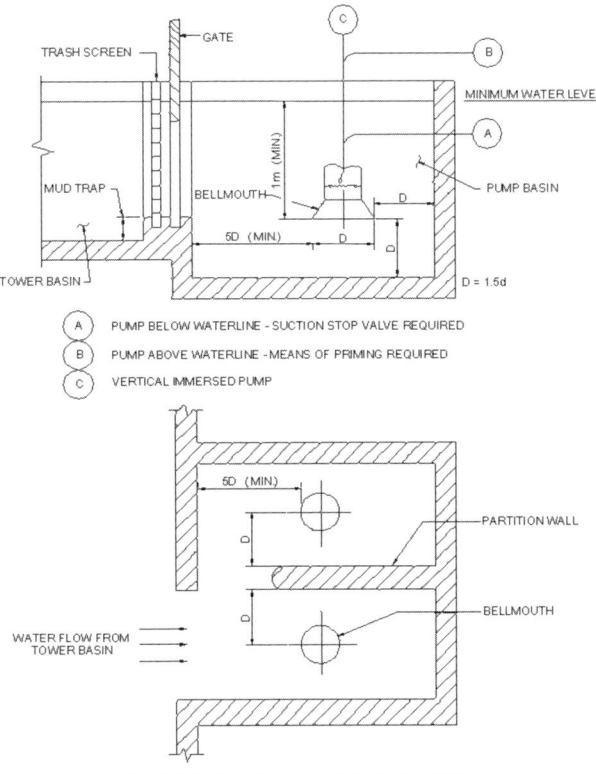

Figure 8–1 *Typical section and plan of a cooling tower pump basin (courtesy of Red Bag/Bentley Systems, Inc.).*

ment of the foundations and a flexible coupling may have to be installed in the suction piping.

Pump Priming

A pump cannot operate without being filled with liquid;, therefore, the minimum design liquid level in the pump basin should be above the casing of the pump. For pumps expected to start up automatically, this method is essential, because there is no danger of the idle pump emptying itself. Vertical immersed pumps also are ideal for automatic standbys.

Where a pump must be above the minimum water level, other means are available for priming the pump, but these should be used with care. These include vent ejectors operated by steam, air, or water; a foot-valve; and a priming feed to the casing from a reliable water source.

Bypass

Within large pumps of high throughput, prolonged operation at too low a flow overheats the pump and water within it and may cause damage to the pump. Therefore, these pumps should be fitted with an open-ended bypass from the discharge line back into the pump basin, terminating below the minimum water level. A means should be provided to prevent high velocity streams from disturbing the flow to the pump suction intakes.

Winterizing

In cold climates, steam injection sometimes is employed in the tower to prevent the pump basins from freezing. A steam header is run around the periphery of the tower basin, above the basin wall, and steam ejected via holes drilled into header, onto the water surface in the basin. Steam also is introduced into the pump basin via a sparger; this prevents the water freezing prior to being pumped into the cooling water system.

Chemical Dosing

To inhibit the growth of algae, reduce scaling in the cooling water system, and adjust the chemical balance of the water, inhibitors are added to the cooling water. If dosing is required, a smaller vendor's water treatment package usually is used, and the chemicals are fed into the pump basin near the sluice gate–trash screen. With chemical dosing, the water treatment vendor's recommendations should be followed.

8.4   Piping Support and Stress Issues

The operating conditions in cooling systems are limited. Piping weight or potential environmental (wind or seismic) forces are to be considered for supporting design. Cooling water systems generally are made in GRE (glass-fiber reinforced epoxy) material. The support and stress issues are best handled in close cooperation with the vendor of the material.

CHAPTER 9

Relief Systems

9.1 Introduction

Relief systems are used to limit the difference between the internal and external equipment pressure. The relief valves prevent damage to mechanical equipment, piping systems, and instruments in case of upset (pressure) conditions in a process plant. Example of relief systems in a process plant are

- Pressure relief valves for protection of static equipment, such as columns, drums, and heat exchangers.
- Thermal relief valves for protection of offsite piping.
- Pressure relief valve pump mechanical casing or mechanical seal protection.
- Pressure relief valves upstream of the first valve on the discharge line of positive displacement pumps.

Since relief systems are a safety item for the plant, the design of the system has to be done with great care. The design of a relief system is governed by standards such as

- API Recommended Practice 520, Design and Installation of Pressure-Relieving Systems in Refineries.
- API Recommended Practice 2000, Venting Atmospheric and Low Pressure Storage Tanks.
- British Factory Acts, in relation to steam raising plants, if applicable.

as well as by the requirements of local authorities and clients and good engineering practice.

The relief and safety systems are important aspects of the process plant design and need to be addressed at an early stage of the design process. Changes to safety systems or area classification can have a large impact on the layout of the plant.

Relief valves may discharging either to atmosphere or into a closed system.

9.2　Types of Relief Devices

The basic types of relief devices are relief valves (pressure or thermal), rupture discs, breather valves, and others, such as open vent valves with a flame arresting device and manhole or gauge hatches that permit the cover to lift.

The pressure and thermal relief valves are safety devices on piping and static equipment. The thermal relief valve usually is found in piping systems or small equipment. The thermal relief valve releases a small amount of liquid to a safe place to reduce the overpressure caused by temperature. Transport pipe lines that can be blocked by valves use thermal relief valves to safeguard the piping from overpressurizing due to temperature raising by sunlight.

The pressure relief valve is used for larger containments, such as vessels and heat exchanges. The pressure relief valve is used not only in liquid but also vapor and gas service. The overpressure caused by upset conditions needs to be handled by releasing a larger amount of vapor or gas to a safe location. For example, if an LPG storage tank faces an overpressure condition, large amounts of LPG need to be released to reduce the pressure.

The rupture disc is a plate that breaks open at a certain set pressure. The disc breaks open only once; if it is open, it needs to be replaced. Rupture discs are found in eroding, corroding, or contaminated processes. The rupture disc commonly is used to protect a safety relief valve from contact with the process fluids.

Breather valves are used in processes where only a low differential pressure is allowed. For example, a storage tank can have only a limited amount of external pressure before the wall shows unacceptable deformation. The storage tank would collapse under a vacuum condition.

9.3　Applicable International Codes

Numerous international codes and standards apply to the various types of tanks and storage vessels used in hydrocarbon processing plants and listed here are several of the most important, along with their scope and table of contents. The design and specifying of these items of equipment are the responsibility of the mechanical engineer; however, a piping engineer or designer will benefit from being aware of these documents and reviewing the appropriate sections that relate directly to piping or a mechanical-piping interface. The standards are

- ANSI/API Standard 521.
- API Standard 526.

9.3.1 ANSI/API STANDARD 521. Pressure-Relieving and Depressuring Systems and ISO 23251 (Identical). Petroleum and Natural Gas Industries—Pressure-Relieving and Depressuring Systems

Scope

This international standard is applicable to pressure-relieving and vapor-depressuring systems. Although intended for use primarily in oil refineries, it also is applicable to petrochemical facilities, gas plants, liquefied natural gas (LNG) facilities, and oil and gas production facilities. The information provided is designed to aid in the selection of the system that is most appropriate for the risks and circumstances involved in various installations. The international standard is intended to supplement the practices set forth in ISO 4126 or API Recommended Practice 520-I for establishing a basis of design.

The international standard specifies requirements and gives guidelines for examining the principal causes of overpressure, determining individual relieving rates, and selecting and designing disposal systems, including such component parts as piping, vessels, flares, and vent stacks. The standard does not apply to direct-fired steam boilers.

Piping information pertinent to pressure-relieving systems is presented in Section 7.3.1.

Table of Contents

API Foreword.

Foreword.

Introduction.

1. Scope.

2. Normative References.

3. Terms and Definitions.

4. Causes of Overpressure.

    4.1. General.

    4.2. Overpressure Protection Philosophy.

    4.3. Potentials for Overpressure.

    4.4. Recommended Minimum Relief System Design Content.

    4.5. List of Items Required in Flare-Header Calculation Documentation.

    4.6. Guidance on Vacuum Relief.

5. Determination of Individual Relieving Rates.

    5.1. Principal Sources of Overpressure.

    5.2. Sources of Overpressure.

    5.3. Effects of Pressure, Temperature, and Composition.

    5.4. Effect of Operator Response.

    5.5. Closed Outlets.

    5.6. Cooling or Reflux Failure.

    5.7. Absorbent Flow Failure.

    5.8. Accumulation of Non-condensables.

    5.9. Entrance of Volatile Material into the System.

    5.10. Failure of Process Stream Automatic Controls.

    5.11. Abnormal Process Heat Input.

    5.12. Internal explosion (Excluding Detonation).

    5.13. Chemical Reaction.

    5.14. Hydraulic Expansion.

    5.15. External Pool Fires.

    5.16. Jet Fires.

    5.17. Opening Manual Valves.

    5.18. Electric Power Failure.

    5.19. Heat-Transfer Equipment Failure.

    5.20. Vapour Depressuring.

    5.21. Special Considerations for Individual Pressure-Relief Devices.

    5.22. Dynamic Simulation.

6. Selection of Disposal Systems.

    6.1. General.

    6.2. Fluid Properties That Influence Design.

    6.3. Atmospheric Discharge.

    6.4. Disposal by Flaring.

6.5. Disposal to a Lower-Pressure System.

6.6. Disposal of Liquids and Condensable Vapours.

7. Disposal Systems.

7.1. Definition of System Design Load.

7.2. System Arrangement.

7.3. Design of Disposal System Components.

7.4. Flare Gas Recovery Systems.

9.3.2 API Standard 526. Flanged Steel Pressure Relief Valves

Scope

This standard is a purchase specification for flanged steel pressure relief valves. Basic requirements are given for direct spring-loaded pressure relief valves and pilot-operated pressure relief valves as follows:

- Orifice designation and area.
- Valve size and pressure rating, inlet and outlet.
- Materials.
- Pressure-temperature limits.
- Center-to-face dimensions inlet and outlet.

For the convenience of the purchaser, a sample specification sheet is given in Appendix A. Nameplate nomenclature and requirements for stamping are detailed in Appendix B.

Table of Contents

1. General.

1.1. Scope.

1.2. Referenced Publications.

1.3. Definitions.

1.4. Responsibility.

1.5. Conflicting Requirements.

1.6. Orifice Areas and Designations.

2. Design.

2.1. General.

2.2. Determination of Effective Orifice Area.

2.3. Valve Selection.

2.4. Dimensions.

2.5. Lifting Levers.

9.4   Inlet and Outlet Piping

9.4.1   Pressure Relief Valve Installation

For general requirements of relief valves installation at vessel or pipework, see Figures 9–1 through 9–8.

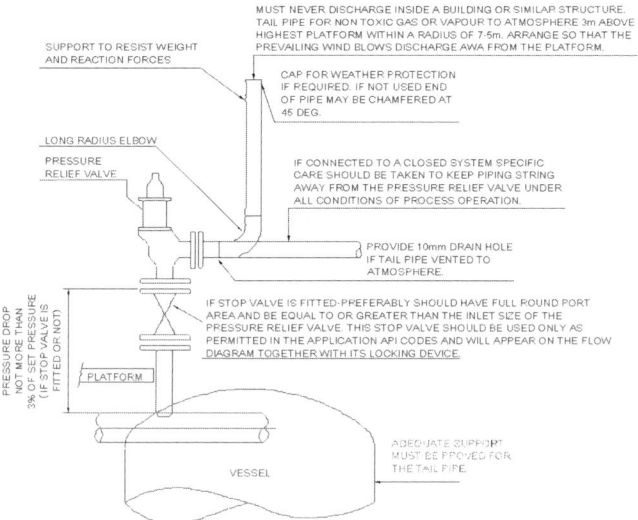

Figure 9–1   *Recommended typical relief valve installation with and without stop valve (based on API Recommended Practice 520).*

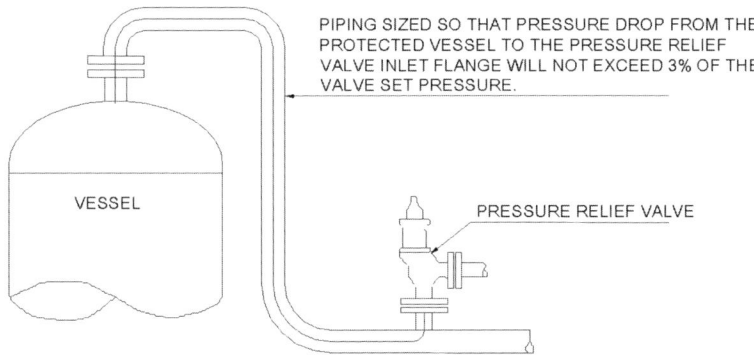

Figure 9–2   *Recommended typical relief valve installation when mounted on the overhead vapor line (based on API Recommended Practice 520).*

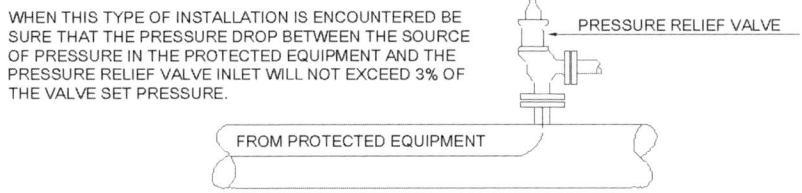

Figure 9–3   *Recommended typical relief valve installation when mounted on the process vapor line (based on API Recommended Practice 520).*

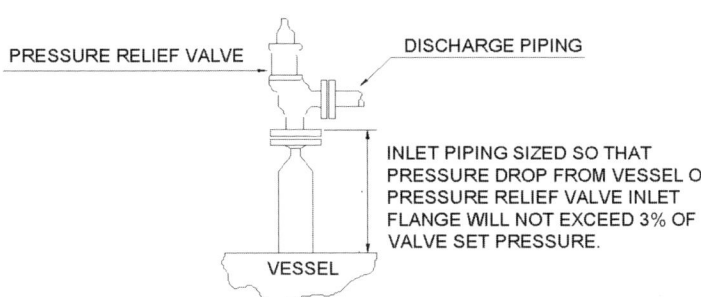

Figure 9–4   *Recommended typical relief valve installation when mounted on long inlet piping (based on API Recommended Practice 520).*

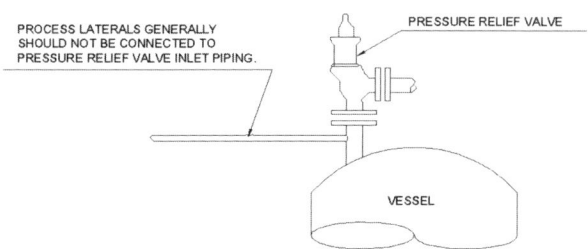

Figure 9–5 *Recommended typical installation to avoid process laterals connected to pressure relief valve inlet piping (based on API Recommended Practice 520).*

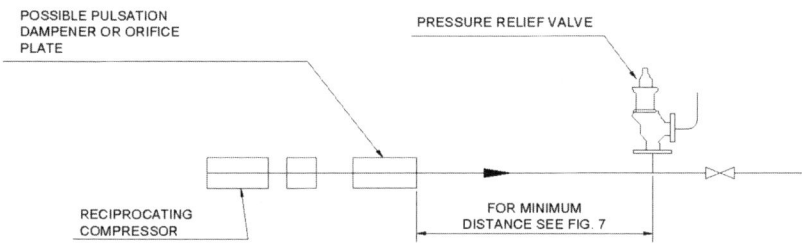

Figure 9–6 *Recommended typical relief valve installation on pulsating compressor discharge (based on API Recommended Practice 520).*

Relief valves should be mounted in a vertical position only.

- For optimum performance, relief valves must be serviced and maintained regularly. To facilitate this, relief valves should be located to enable easy access and removal. Sufficient platform space must be provided around the valve.

- The provision of a lifting device should be considered for large relief valves. For typical installation of relief valves and piping, see Figures 9–1 through 9–8.

- The system inlet and outlet piping should be designed to provide for proper valve performance. To this end, a complete relief system isometric showing the inlet and discharge piping of all relief valves included in the system should be produced. The figures show part of such a system, overall dimensions and line sizes should be included.

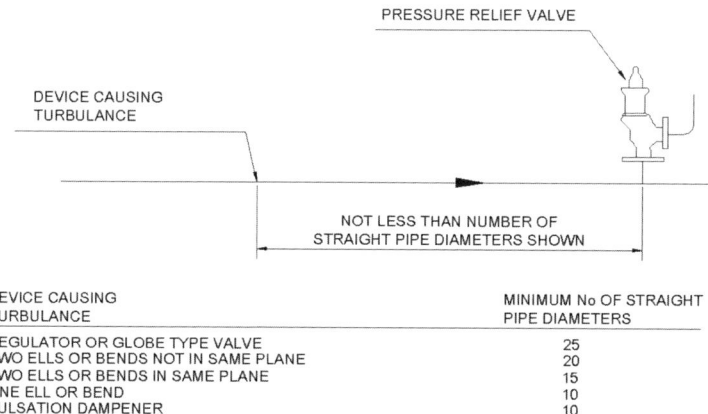

DEVICE CAUSING TURBULANCE	MINIMUM No OF STRAIGHT PIPE DIAMETERS
REGULATOR OR GLOBE TYPE VALVE	25
TWO ELLS OR BENDS NOT IN SAME PLANE	20
TWO ELLS OR BENDS IN SAME PLANE	15
ONE ELL OR BEND	10
PULSATION DAMPENER	10

Figure 9–7 *Recommended typical installation to avoid excessive turbulence at the entrance to the pressure relief valve lateral when mounted on a line (based on API Recommended Practice 520).*

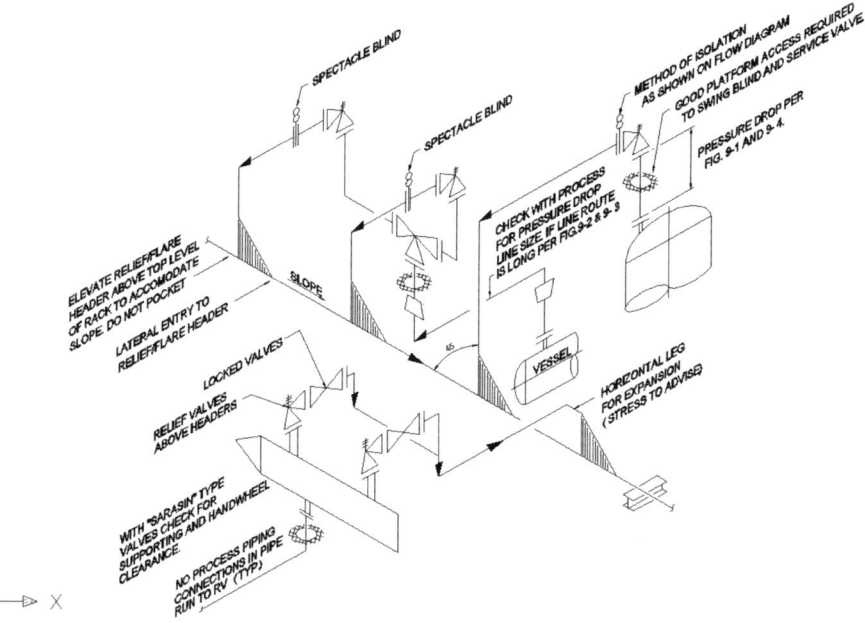

Figure 9–8 *Recommended typical relief valve installation (based on API Recommended Practice 520).*

- The system must first be approved by the Process Department for pressure drop, then by the Stress Section for system flexibility. Should the piping be rerouted due to stress requirements, the Process Department must reapprove the system.

- The inlet piping to a relief valve should be designed so that the pressure drop does not exceed 3% of the relief valve set pressure, see Figures 9–1, 9–2, and 9–4.

- The most desirable installation is that in which the nominal size of the inlet branch piping is the same as or greater than the nominal size of the valve inlet flange and the length does not exceed the face-to-face dimensions of a standard tee and weld neck flange of the required pressure class. The configuration shown on the flow diagram always must be followed.

- Relief valve systems designed to discharge against a constant pressure cannot tolerate a back pressure greater than 10% of the set pressure. Balanced bellows or Balanseal relief valves, which operate practically independent of the back pressure, tolerate a much higher figure; but generally, the higher is the back pressure, the lower the capacity of the relief valve. Due to this, the discharge piping should be kept as direct as possible, see Figures 9–6 and 9–7.

- The flow diagram determines if the discharge is to the atmosphere or a closed discharge header, see below figures for details.

- It is poor practice to mount the relief valve at the end of a long, horizontal inlet pipe through which there is normally no flow.

- Foreign matter may accumulate or liquid may be trapped and may interfere with the operation of the valve or be the cause of more frequent valve maintenance.

- Process laterals generally should not be connected to relief valve inlet piping, see below Figure 9–4.

- Proximity of other valves and equipment, the recommendations laid down in below Figures 9–4, 9–6 and 9–7 should be followed if possible for the minimum number of straight pipe diameters between the device causing turbulence and the relief valve. The above does not apply to relief valves fitted with stop valves, see Figures 9–1 and 9–4.

- Relief valves discharging into a relief header must be located at an elevation above the relief header so that the discharge line is free draining. Under no circumstances may the discharge line or the relief header be pocketed.

9.4.2 Thermal Relief Valve Installation

- Thermal relief valves are provided on cooling services, where a system can be locked in by isolating valves.
- The discharge from the relief valve should be piped to grade in offsite areas and to the nearest drain in process areas; follow details as shown on the flow diagram. See Figure 9–9 for further details.

9.4.3 Rupture Disc Installation

- A rupture disc device is a sacrificial, non-reclosing pressure relief device actuated by inlet static pressure and designed to function by the bursting of a pressure containing disc. The disc may be made of metal or carbon graphite and housed in a suitable holder.
- The purpose of installing a rupture disc upstream of a relief valve is to minimize the loss by leakage through the valve of valuable, noxious, or otherwise hazardous materials or to prevent corrosive gases from reaching the relief valve internals, see Figure 9–10. Typical bursting disc installation at a relief valve.
- The installation of a rupture disc may be called for downstream of a relief valve. In this case, its purpose is to prevent corrosive gases from a common discharge line reaching the relief valve internals.
- Rupture discs may also be installed as a sole relief device, see Figure 9–11.
- Rupture discs always must be installed as shown on the vendor's drawing. For an example of a concave installation see Figure 9–11 and a convex installation see Figure 9–10.

9.4.4 Breather Valve Installation

The venting requirements are determined based on the following:

- In-breathing results from maximum outflow of oil from tanks or from contraction of vapors caused by a maximum decrease in atmospheric temperature. A vacuum-breaking device is fitted directly to the line or vessel to prevent its collapse when the internal pressure falls below atmospheric. If the admission of oxygen into the process is hazardous, a inert gas supply is connected to the vacuum breaker via a pressure reducing valve. Follow the system defined on the flow diagram.

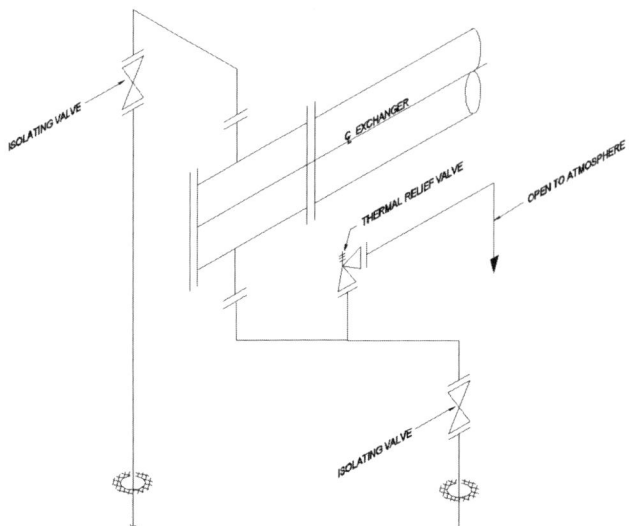

Figure 9–9   *Recommended typical thermal relief valve installation (based on API Recommended Practice 520).*

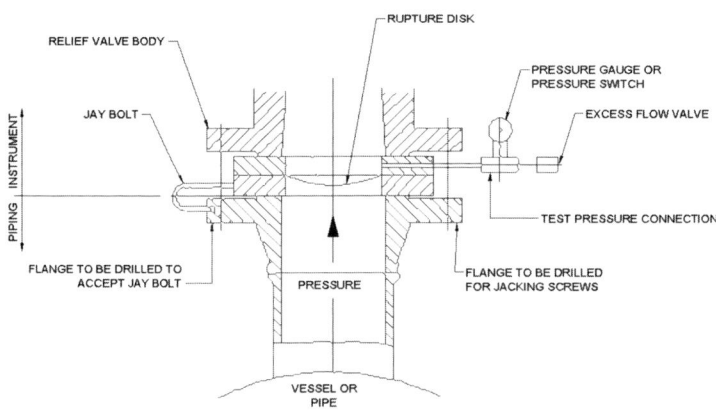

Figure 9–10   *Rupture disc installed to protect relief valve; reverse-buckling disc with the pressure loading on the convex side of the disc (based on API Recommended Practice 520).*

- Out-breathing results from maximum inflow of oil into tanks, which results from a maximum increase in atmospheric temperature (thermal breathing). The devices may be specified on the flow diagram.

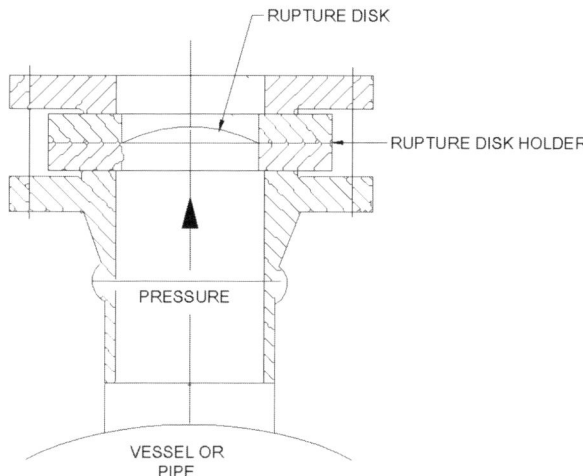

Figure 9–11   *Rupture disc installed as a sole relief device; pressure load on the concave side of the disc (based on API Recommended Practice 520).*

Pressure and vacuum breathing valves are designed to prevent evaporation from storage tanks but to allow breathing when the pressure or vacuum exceeds that specified. For details, see Figure 9–12.

9.5   Piping Support and Stress Issues

- Relief valve inlet and outlet piping should be designed to accommodate forces due to discharge forces.
- The support design of the piping should be calculated for the dynamic effects of the discharge flow.
- Relief system also should be designed to accommodate the thermal effects of the discharge, such as autorefrigeration (by expanding gases) and heat from a high-temperature release.
- The piping system requires adequate expansion loops to allow for thermal expansion. Expansion bellows are not permitted in relief systems.
- The support design requires accommodating the largest forces due to slug flow, water hammer, or other dynamic effects of the discharge flow.

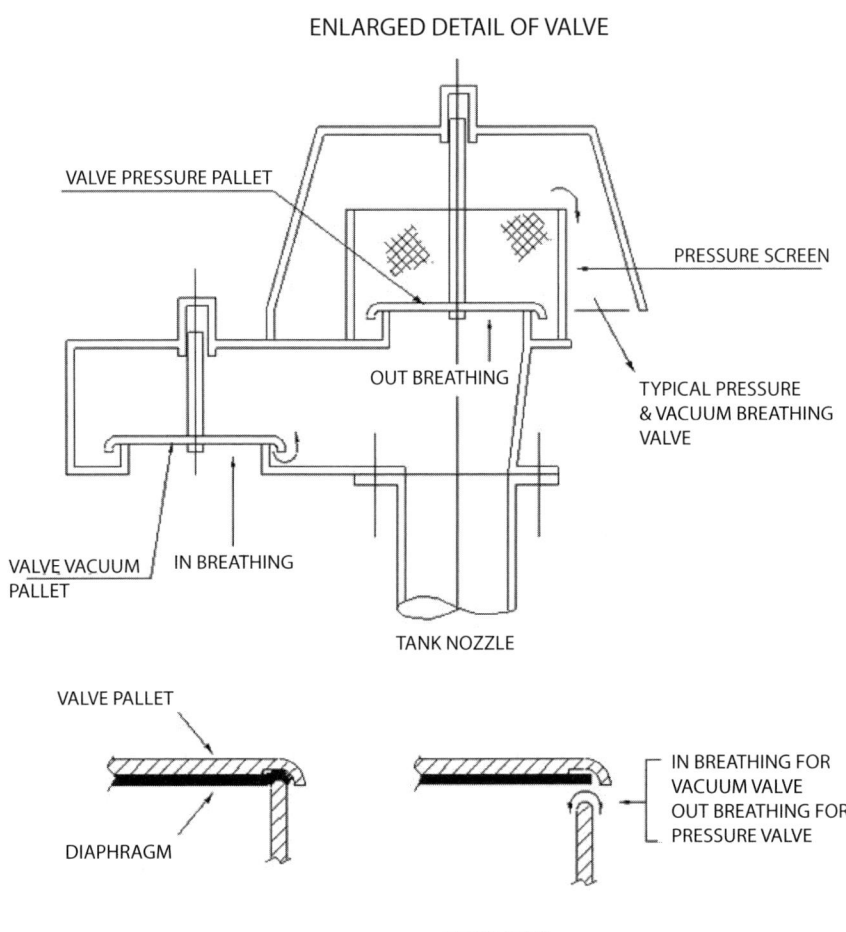

Figure 9–12   *Typical pressure and vacuum breathing valve (based on API Recommended Practice 520).*

CHAPTER 10

Pipe Ways

10.1 Introduction

The pipe way conveys all the main process lines connecting distant pieces of equipment, relief and blowdown headers, all lines leaving and entering the plant, and utility lines supplying steam, air, cooling water, and inert gas to the plant. Electrical and instrument cable trays usually are routed in the pipe way. Pipe ways are classified by their relative elevation to grade.

Pipe way design is a complex activity, and it requires a great deal of information to be completed successfully. Issues that need to be considered are

- Battery limits, valve location, and the isolation philosophy using spade and spacers.
- Catwalk, platform, and ladder access to valves and relief valves in the pipe rack.
- Minimum headroom and clearances below overhead piping or supporting steel within areas.
- Pipe ways and secondary access ways.
- Main access roads.
- Railroads.
- Standards for the minimum spacing of lines in pipe racks.
- Handling and headroom requirements for equipment positioned under pipe racks.
- Operating and safety requirements affecting pipe rack and structure design.
- Location of cooling water lines underground or aboveground.

- Trenched piping, if any, and schematic diagrams such as

  - Process flow diagrams, which show main process lines and lines interconnecting process equipment.

  - Engineering flow diagrams, which are developed from process flow diagrams and show pipe sizes, classes, and line numbers; valves; manifolds; all instruments; and equipment and lines requiring services (water, steam, air, nitrogen, etc.).

  - Utility flow diagrams, which show the required services such as steam, condensate, water, air, gas, and any additional services peculiar to the plant being worked on, such as caustic, acid, and refrigeration lines.

10.2　Types of Pipe Ways

The types of pipe ways are pipe racks, pipe tracks, trenched piping, and underground piping.

10.2.1　Pipe Racks

Pipe racks are used for routing overhead piping, which is supported on steel or concrete bents.

10.2.2　Pipe Tracks

A pipe track is used for aboveground piping supported on concrete sleepers at grade level. Pipe tracks are used mostly for offsite areas where equipment is well spaced out.

10.2.3　Trenched Piping

Trenched piping is below-ground piping laid in a network of inter-connected trenches. The design of the trenches usually is the responsibility of the civil engineers. Trenching pipework is costly and usually undesirable, unless trenches are wide, shallow, and well vented; heavy gases may settle and create a fire hazard through the length of the trench. For these reasons, only pump-out lines, chemical sewers, or chemical drain collection systems sometimes are placed in trenches and routed to a pit or underground collection tank.

10.2.4　Underground Piping

Underground piping is piping directly buried below ground level. The groundwork for the underground piping is also the responsibility of the Civil Engineering Department. Due to costly maintenance and the usually corrosive nature of soil, this method of routing generally is reserved for sewers, fire water mains, and drain lines. In some plants, especially in cold climates, cooling water lines are buried below the frost line to avoid freezing. This option should be determined at the beginning of a job and is generally a customer requirement.

10.3 Piping and Support

10.3.1 Initial Design

During the initial design phase, the following points must be considered to establish preliminary routing:

- Use the plot plan and process flow diagrams to make a preliminary assessment of which portion of process lines will be located in a pipe rack and which lines will interconnect directly to nozzles on adjacent items of equipment.
- Draw lines to be located on pipe racks on the print of the plot plan.
- Some idea of the utility piping required must be established and included.
- Coordination with Instrument and Electrical Section is essential to assess what additional rack space might be required to accommodate cable trays. This action provides a preliminary visual idea of the pipe-rack space required.

10.3.2 Design Development

As the project develops and with the receipt of engineering flow diagrams and utility flow diagrams, a more complete and accurate assessment of rack space is possible. Utility headers generally run the whole length of the pipe rack, so they should be taken into account when estimating additional space required. To assist the Process Department in sizing utility headers in the pipe way, a line, routing on a copy of the plot plan, showing order of takeoffs is required.

Certain types of piping and pipe-way items require special consideration: process lines, relief headers, and instrument and electrical cable trays.

Process Lines

Generally, lines that are interconnections between nozzles on pieces of process equipment more than 6 m apart should be run on a pipe rack. Equipment that is closer may connect directly inside piping areas:

- Product lines that run from vessels, exchangers, or pumps to battery or unit limits.
- Crude or other charge lines entering the unit that run along a pipe rack before connecting to process equipment, furnaces, exchangers, holding drums, or booster pumps.

Relief Headers

Individual relief lines, blowdown lines, and flare lines should be self-draining from all relief valve outlets to knockout drums, flare stacks, or to a point at the plant limit. To achieve this, lines connect into the top of the header, usually at 45° in direction of flow. To eliminate pockets and obtain the required slope to a knockout drum, some relief headers are placed on the highest elevation of the main pipe rack, or on a T–support, which is an extension of a pipe-rack column.

Instrument and Electrical Cable Trays

Often instrument and electrical cable trays are supported on the pipe-rack track. Space must be allocated to accommodate them at the initial design stage. Due to the possibility of induced current interference, instrument and communication cable trays must be located away from electrical and power cable trays. Consult with Instrument/Electrical Department to determine separation requirements.

10.3.3 Pipe-Rack Width

The width of a pipe rack is influenced by the number of lines, electrical or instrument cable trays, and the space for future lines. The width of a pipe rack may be calculated using the following method: First estimate number of lines as described; add the number of lines up to 18 in diameter in the most dense section of the pipe rack; then an estimate of the total width in meters (W) is

$$W = (f \times N \times S) + A \text{ m,}$$

where

f, the safety factor = 1.5 if the lines have been laid out as described in initial evaluation, and $f = 1.2$ if the lines have been laid out as described under development;

N = the number of lines less than 18 diameter;

S = average estimated spacing between lines in mm;

Usually, $-S = 300$ mm, but $-S = 230$ mm if lines in the pipe rack are smaller than 10;

A = additional width required meters for lines larger than 18, future lines, instrument and electrical cable trays, and any slot for pump discharge lines 500 mm–1 m.

Thus the total width is obtained. If W is bigger than 9 m, usually two pipe rack levels are required. Note: At the beginning of a job, W usually should include 30–40% of clear space for future lines.

The width of the pipe rack may be increased or determined by the space requirement and access to equipment arranged under the pipe rack. Figure 10–1 shows typical pipe racks bents with tabulated dimensions. The total available pipe-rack width of each type of support is in Table 10–1. This table can be used for selection. The most commonly used pipe-rack supports are types 2, 3, 4, and 5.

Spacing between Pipe-Rack Bends

The term, *bent* refers to the vertical structural member, which can be constructed of steel or concrete. Normal spacing between pipe-rack bents varies between 4.6 m and 6 m. This may be increased to a maximum of 8 m. Consideration must be given to

- Smaller lines, which must be supported more frequently.
- Liquid-filled lines, requiring shorter span than gas- filled lines.
- Hot lines, which span shorter distances than cold lines of the same size and wall thickness.
- Insulated lines, small-bore, cold-insulated lines must be supported at relatively short intervals due to weight of insulation.
- Space requirements of equipment at grade sometimes influence pipe-rack bent spacing.

10.3.4 Pipe-Rack Elevation

Pipe-rack elevation is determined by the highest requirement of the following:

- Headroom over the main road.
- Headroom for access to equipment under the pipe rack.
- Headroom under lines connecting the pipe rack and equipment located outside.

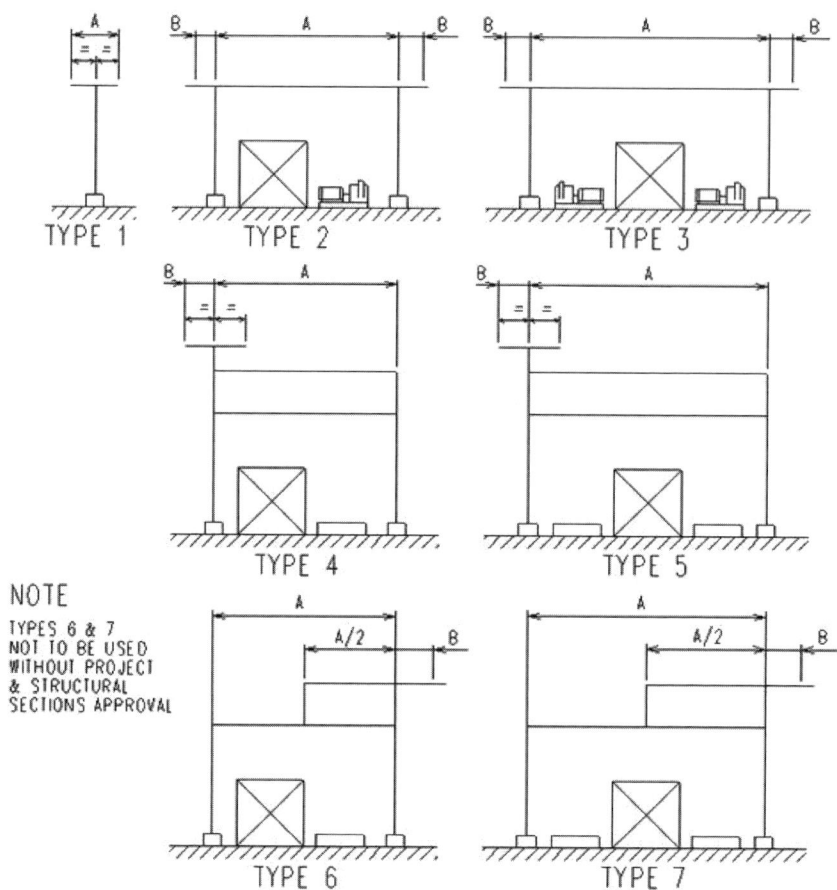

Figure 10–1   *Typical pipe rank bents (courtesy of Red Bag/Bentley Systems, Inc.).*

The size of the steel or concrete beam supporting overhead piping must be taken into consideration.

Elevation at Pipe-Rack Intersection

Where two two-tier pipe racks meet, it is essential that elevations of lateral pipe racks slot between elevations of main pipe rack. Figure 10–2 (top) illustrates this requirement. The choice of top elevation of the lateral pipe rack midway between the top and bottom main pipe-rack elevation allows turning up or down at the intersection. Generally, lines running at right angles to main pipe rack are assigned elevations 500 mm–1 m higher or lower (depending on headroom

Table 10–1   Pipe-Rack Space

Type No.	Total Available Width, W, in mm		Pipe Rack Width, A	Cantilever Width, B	Number of Elevations
	Without Cantilever	With Cantilever			
1	3000	—	3000	—	1
2	6,000–7,300	9,150–10,400	6,000–7,300	1,500	1
3	8,500–9,750	11,600–12,800	8,500– 9,750	1,500	1
4	11,900–14,300	13,700–16,150	6,100–7,300	900–1,200	2
5	16,800–19,200	18,600–21,000	8,500–9,750	900–1,200	2
6	8,500–10,400	11,000 –12,800	6,100–7,300	900–1,500	1.5
7	12,200–13,400	14,650–15,850	8,500–9,750	900–1,500	1.5

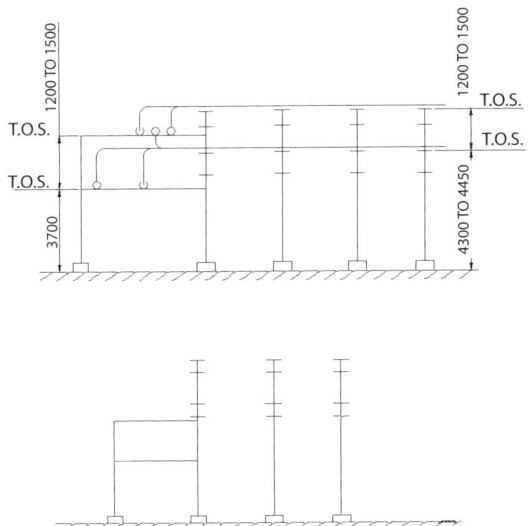

Figure 10–2   *Typical pipe rank intersection (courtesy of Red Bag/Bentley Systems, Inc.).*

requirements) than lines running in main pipe rack. The 500 mm differential between pipe runs is the absolute minimum.

Figure 10–2B shows a pipe-rack intersection where the respective main and lateral pipe-rack elevations do not slot between each other. This design complicates routing of lines from one pipe rack to another, especially where lines run on the bottom levels of both pipe racks. Avoid this design at all cost.

Where a single-tier pipe rack turns 90° and all lines can be kept in the same sequence in both directions, no elevation difference is necessary. When the line sequence changes, introduce an elevation change at the turn (see Figure 10–3).

10.3.5 Line Location in Pipe Racks

One-Tier Pipe Racks

Figure 10–4 shows a cross section of a single-level piperack. Heavy pipes that are either of large diameter, heavy wall thickness, or transporting liquid, regardless of service, are placed over or near the pipe-rack columns. This simplifies steelwork or concrete pipe rack design. Centrally loaded columns and reduced bending moment on the beam result in a lighter overall design. Place process and relief lines next to these heavy pipes, lines serving the left-hand areas of plant on the left, lines serving the right-hand areas on the right.

The central pipe-rack portion generally is reserved for utility lines, which may serve both the right- and left-hand areas on the plant. However, utility lines serving one or two pieces of specific equipment should be on the same side of the pipe rack as the equipment to which they connect.

Process lines that connect equipment on both sides of the pipe rack should be placed close to utility lines and can be on either side of pipe rack, depending on the location of the equipment they serve. The position of product lines is influenced by their routing after leaving the unit, right- (or left-) turning lines should be on the right- (or left-) hand side of the pipe rack.

If possible, a centrally placed section of the pipe rack is reserved for future lines. This section should be clear for the complete length of the pipe rack. If this is impracticable, then a series of smaller sections, running the whole length of the pipe rack, are to be provided (see Figure 10–5).

Two-Tier Pipe Racks

Where the number of lines dictates the use of a two-level pipe rack, the utility lines are placed on the top level and the process lines on

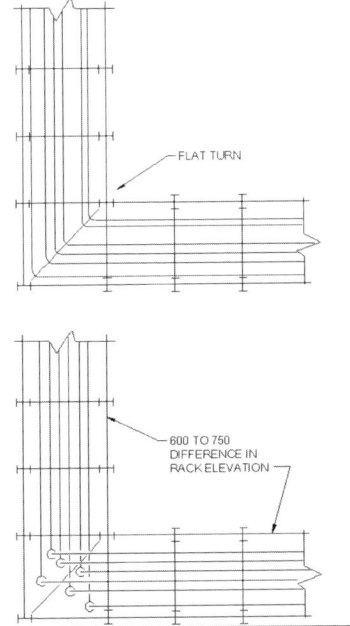

Figure 10–3   *Single-tier rack turning 90° (courtesy of Red Bag/Bentley Systems, Inc.).*

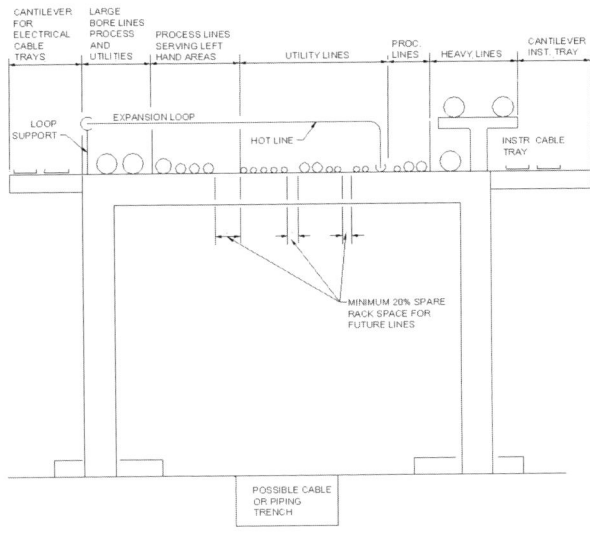

Figure 10–4   *Cross section of a single-level pipe rack (courtesy of Red Bag/Bentley Systems, Inc.).*

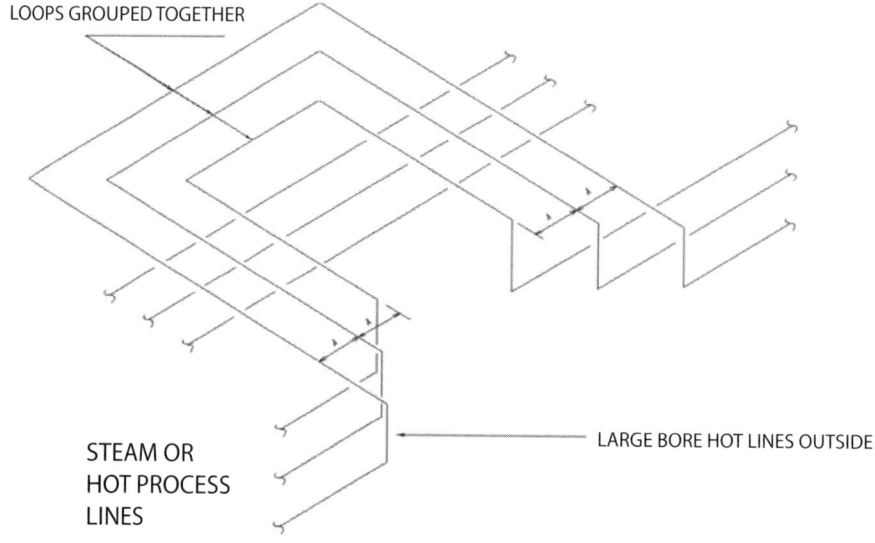

LOOPS GROUPED TOGETHER

STEAM OR
HOT PROCESS
LINES

LARGE BORE HOT LINES OUTSIDE

Figure 10–5   *Expansion loops (courtesy of Red Bag/Bentley Systems, Inc.).*

the bottom level. This is not a rigid rule, and where piping economy dictates, certain process lines may be routed on the top level. The line sequence arrangement should follow the philosophy outlined previously.

A good rule of thumb is to allow three times the largest NPS size between the top of steel (TOS) of the tiers and one and a half times the largest NPS size between the strainers and the tier TOS.

Position of Hot Lines

It is advantageous for pipe supports to group together hot lines requiring expansion loops, preferably on side of the pipe rack. Horizontally elevated loops over the pipe rack are commonly used to minimize the effects of expansion of the hot lines, the hottest and largest line being on the outside (see Figures 10–6 and 10–7).

Line Spacing

For line spacing, use the recommended pipe-rack spacing chart as per job specification. Note that, in certain cases, it is necessary to deviate from the standards quoted previously. At point A, due to possible differential expansion, line spacing may have to be increased to allow for movement of lines at startup. To determine the expansion of the hot lines, use the client engineering guide for

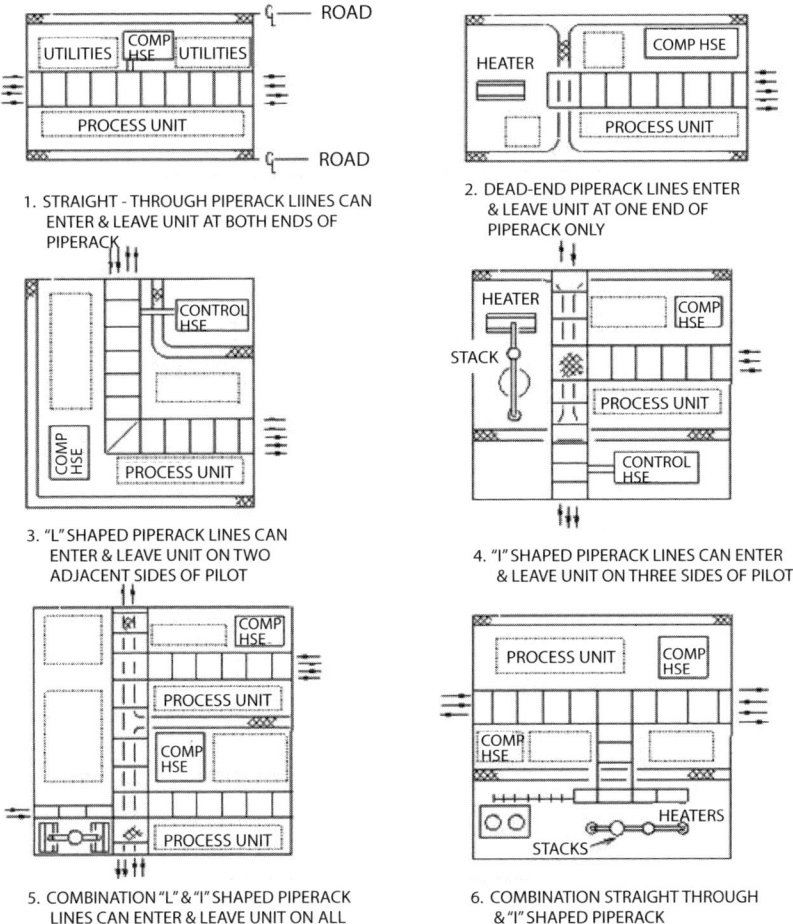

Figure 10–6 *Pipe rack layouts (courtesy of Red Bag/Bentley Systems, Inc.). Note: It is obvious from these examples that a complex piping arrangement can be broken into a combination of simple arrangements.*

"thermal expansion of pipe materials." An example is shown in Table 10–2.

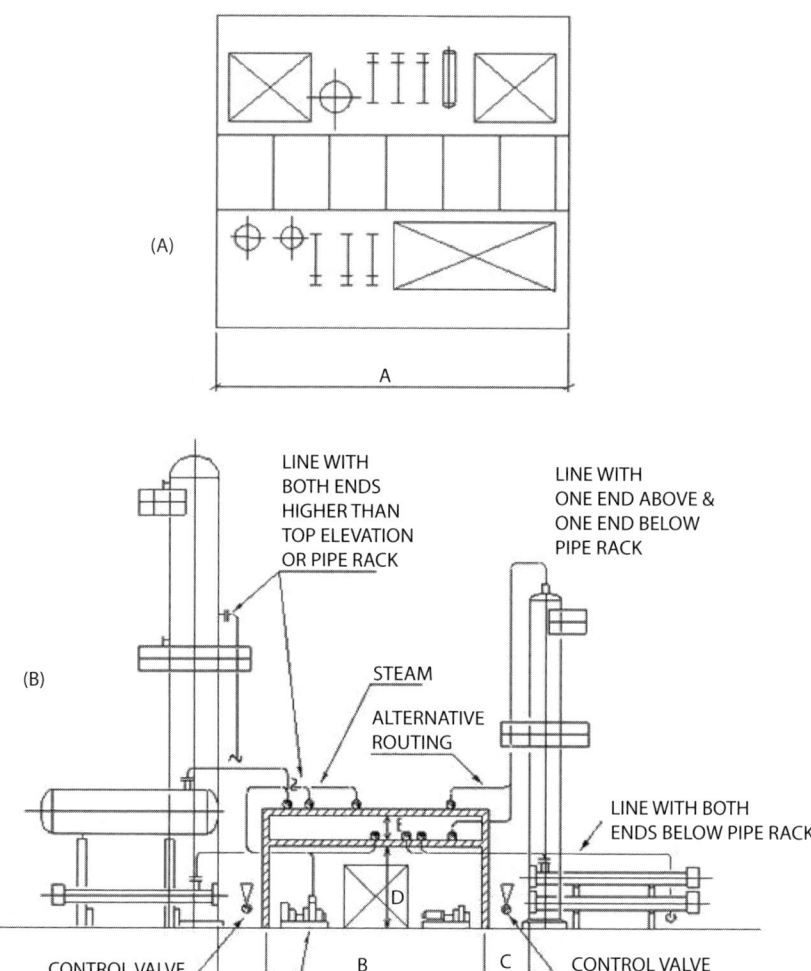

Figure 10–7   *Pipe rack layout broken into simple arrangements (courtesy of Red Bag/Bentley Systems, Inc.).*

Table 10–2 Linear Thermal Expansion between 70°F (21°C) and Indicated Temperature in in./100 ft (mm/30.48m)

Temp. (°F)	Temp. (°C)	Carbon Steel, Low Chrome through 3 Cr Mo		5 Cr Mo through 9 Cr Mo		Austenitic Stainless Steel	
		in.	mm	in.	mm	in.	mm
–50	–46	–0.84	–21	–0.79	–20	–1.24	–32
–25	–32	–0.68	–17	–0.63	–16	–0.98	–25
0	–18	–0.49	–12	–0.46	–12	–0.72	–18
25	–4	–0.32	–8	–0.30	–8	–0.46	–12
50	10	–0.14	–4	–0.13	–3	–0.21	–5
70	21	0.00	0.00	0.00	0.00	0.00	0.00
100	38	0.23	6	0.22	6	0.34	9
125	52	0.42	11	0.40	10	0.62	16
150	66	0.61	15	0.58	15	0.90	23
175	80	0.80	20	0.76	19	1.18	30
200	93	0.99	25	0.94	24	1.46	37
225	107	1.21	31	1.13	29	1.75	44
250	121	1.40	36	1.33	34	2.03	52
275	135	1.61	41	1.52	39	2.32	59
300	149	1.82	46	1.71	43	2.61	66
325	163	2.04	52	1.90	48	2.90	74
350	177	2.26	57	2.10	53	3.20	81
375	191	2.48	63	2.30	58	3.50	89
400	204	2.70	69	2.50	64	3.80	97
425	219	2.93	74	2.72	69	4.10	104
450	232	3.16	80	2.93	74	4.41	112

Table 10–2   Linear Thermal Expansion between 70°F (21°C) and Indicated Temperature in in./100 ft (mm/30.48m) (cont'd)

Temp. (°F)	Temp. (°C)	Carbon Steel, Low Chrome through 3 Cr Mo		5 Cr Mo through 9 Cr Mo		Austenitic Stainless Steel	
		in.	mm	in.	mm	in.	mm
475	246	3.39	86	3.14	80	4.71	120
500	260	3.62	92	3.35	85	5.01	127
525	274	3.86	98	3.58	91	5.31	135
550	288	4.11	104	3.80	97	5.62	143
575	302	4.35	110	4.02	102	5.93	151
600	316	4.60	117	4.24	108	6.24	159
625	330	4.86	123	4.47	114	6.55	166
650	343	5.11	130	4.69	119	6.87	175

Note: Identify the pipe material and the design temperature (use the higher number for in-between temperatures), then apply the following formula to calculate the expansion:

Length of pipe in feet × times expansion per 100 ft = expansion in inches

Length of pipe in meters × expansion per 30.48 m = expansion in millimeters.

10.3.6 Piping Economy in Pipe Racks and Its Influence on Plant Layout

Pipe-Rack Layout

- The plant layout determines the main pipe-rack piping runs.
- The shape of the pipe rack is the result of plant arrangements, site conditions, client requirements, and overall plant economy.
- Piping economy depends primarily on the length of lines routed in the pipe-rack. The figures in this chapter show critical dimensions that influence overall cost. These dimensions depend on overall plant layout and should be carefully considered when the plot is arranged.

- • Dimension A, (see Figure 10–1) is the total length of pipe rack and is governed by the number and size of equipment, structures, and buildings arranged along both sides of the pipe rack. On average, 3 m of pipe-rack length ise required per item of process equipment, a good layout can reduce the pipe-rack length and, thus, its cost.

- • Equipment in pairs, stacked exchangers supported from towers, two vessels combined into one, closely located towers with common platforms, process equipment located under the piperack—these are examples that help shorten pipe-rack length.

- • In a well-arranged plant, the average length of a pipe-rack per item of process equipment can be reduced to 2.1–2.4 m.

- • Careful selection of dimensions B and C, minimizes connection equipment on opposite sides of the pipe rack. Dimension C normally is no more than 1.8–3 m.

- • Overgenerous dimensioning in dimensions D and E increase vertical pipe lengths.

- • Maximize use of available platforms for access to valves.

- • Where air fins are located above the pipe rack, use associated air fin maintenance platforms, modifying their extent, if necessary. This method is cheaper than adding special platforms in the pipe rack.

10.3.7 Pipe-Rack General Arrangement Checklist

Critically review the pipe-rack layout against the latest information from all disciplines. Some lines may require rerouting for maximum piping economy. Check the loops required for nesting. As many loops as possible should be combined in a loop bay, having due regard to stress requirements.

Vibrating Lines

The general check items include

- • Avoid changes in direction.
- • Avoid long overhanging bends without support.
- • Use large radius bends where possible (check with job specification).
- • Avoid T-connections as far as possible, flow should enter along the run of the T-connection and seldom in the branch.

Low-point pockets are to be avoided in the following lines:

- Steam (trap any pockets and dead ends).
- Slurry.
- Blowdown (these lines must be self-draining).
- Caustic, acid, and phenol (all these services to be self-draining).
- Relief valves, both inlet and outlet.
- Vapor to knockout pots.
- Heavy products, bitumen, wax.
- Pump suctions.
- Lethal and toxic substances.

High-point pockets are to be avoided in the following lines:

- Pump suctions.
- Light ends.
- Vapor-liquid mixes (hot tower bottoms, reflux lines).
- Crude lines.

On hot lines, check shoe requirements and clearances at changes of direction (pipe expansion). Provide vents at high points and drains at low points. Provide steam traps at low points, upstream of loops and dead ends, and via condensate drip legs.

On steam, air, and condensate headers, takeoffs are to be from the top of the headers. Relief-valve headers may have high or low elevation. Before finalizing the elevation of a relief-valve header, consider the elevation of all relief-valve discharges and the knockout drum at flare.

Bends, if used (check with job specification) require consideration of

- Where lines change elevation, bends may be used, providing difference in elevation is adequate and specification permits.
- For header takeoffs in pipe racks, use elbows.
- Special piping:
  - Catalyst lines, 5 diameters of the pipe (minimum).
  - Vibrating piping, 5 diameters.
  - Small bore usually below 2 (or client preference).

Pipe setting is to be avoided in large bore lines. Small bore lines to be set only where absolutely necessary.

The Pipe Stress Department will advise which piping systems require formal stress analysis.

Also, consider the following:

- In specifying supports, avoid long unsupported overhangs.
- Check steelwork clearances for addition of fireproofing (lower elevation of pipe rack), brackets, gussets, and thermal expansion of lines.
- Check concrete support clearances for local thickening of concrete columns due to method of fabrication adopted (i.e., corbels).
- Check clearances and accessibility of valves. Make full use of extending platforms for operation (i.e., air fin maintenance platforms).
- Preferably do not use chain wheels; however, if necessary, check chain clearances. Check for accessibility of the spading and valving at battery limit; if necessary, provide an access platform.

10.3.8 Pipe Tracks

This type of pipe way generally is associated with offsite areas, where equipment is well spaced out and land space is not at a premium.

Pipe-Track Width

The pipe-track width may be estimated using the method detailed previously for pipe-racks.

Spacing of Pipe-Track Sleepers

Pipe-track sleepers are relatively cheap; therefore, piping economy is dictated by the recommended span of the smallest line in the track.

Where small bore lines are limited, sleeper spacing may be determined by the pipe span of large bore lines, provided small bore lines are supported off the larger lines at adequate intervals. An angle with U-bolts is sufficient (check with the Pipe Support Section).

For recommended pipe support spans, use the client standard: On an average minimum span = 3 m; maximum span = 6 m. This depends on the line size and media carried in pipes (i.e., gas or liquid).

All lines must be supported. At changes of direction due to long overhangs, for narrow pipe tracks, a diagonal corner sleeper is recommended; on wide pipetracks, use the alternative method of two short sleepers located near corner (see Figure 10–8).

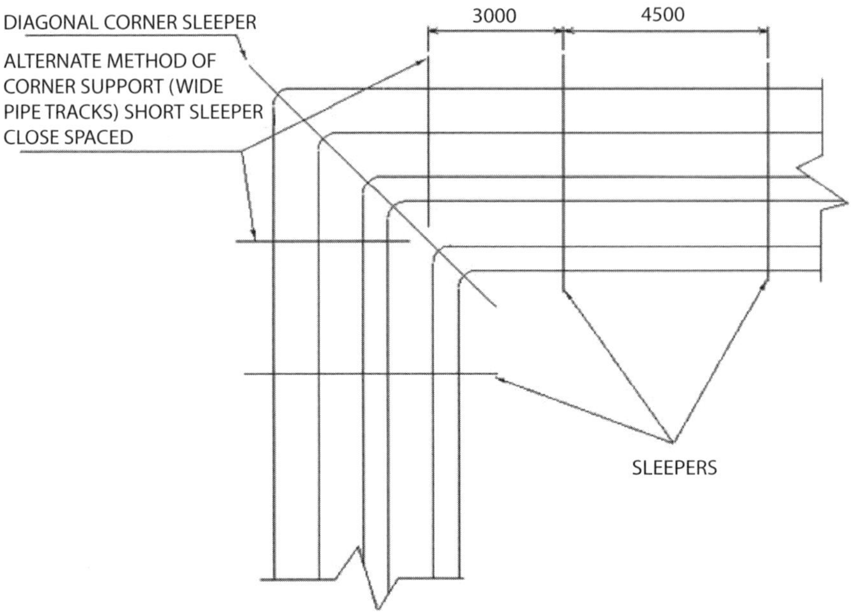

Figure 10–8   *Spacing pipe track sleepers (courtesy of Red Bag/Bentley Systems, Inc.).*

Pipe-Track Elevation

Pipe-track elevation is set by maintenance access to piping items located underneath the pipe track,; that is, drains and steam traps. A minimum of 12/300 mm clearance between underneath of lines and grade is recommended; where necessary, this may be increased to 18/450 mm.

As pipe tracks generally are single tier, no change in elevation is necessary at changes of direction. This is effected by use of a flat turn (see Figure 10–2).

Line Location

Line location with reference to bore and weight is unnecessary, as all pipes are supported on sleepers, which rest directly on the ground. Line routing is all important. All lines interconnecting process equipment or storage tanks located on left-hand side of the pipe track are placed on the left-hand side. Similarly, all lines interconnecting equipment located on right-hand side of the pipe track are placed to the right of pipe track. Lines connecting equipment located on either side of the pipe track are placed near the center of pipe track.

Road Crossings

The standard method is to provide culverts under access roads. Elevating piping on a crossover rack is expensive and introduces unnecessary pockets in the lines thus routed. When determining the width and height of a culvert, care must be taken to allow sufficient room round the pipework for maintenance, insulation, and painting. Where only one or two lines cross a road, crossing may be by means of sleeves set under the roadway.

Access Ways

In areas needing frequent access, platforms may be provided across the pipe track.

Valves

Where possible, valves should be grouped at the edge of an access platform for ease of operation. Drain valves, where possible, should be brought to outside of pipe track for ease of operation. The same applies to steam trap assemblies.

10.3.9 Expansion Loops

Horizontal expansion loops elevated above the pipe track should be provided where necessary. Group all hot lines requiring expansion loops with the hottest and largest line on the outside, on one side of pipe track. (Generally, the side chosen is that which has the highest number of takeoffs serving equipment on that side.)

10.4 Trenched Piping

In most plants, trenches are avoided due to the problems associated with this type of pipe way: high initial cost and fire hazard. Where trenchers are used, they are to route lines such as pump-out lines, chemical sewers, and chemical drains.

Trenches must be deep enough and wide enough to allow sufficient clearance between the trench wall and piping. The minimum acceptable clearance is 150 mm between the outside of pipe and the inside of wall. This allows for installation of piping, painting, and future maintenance (see Figure 10–9). The total required width of the trench may be determined using the method just detailed. The Piping Department will advise Civil Department of the requirements.

The line location in a trench carrying a number of lines should be carefully chosen for maximum piping economy consistent with the stress requirements, if any.

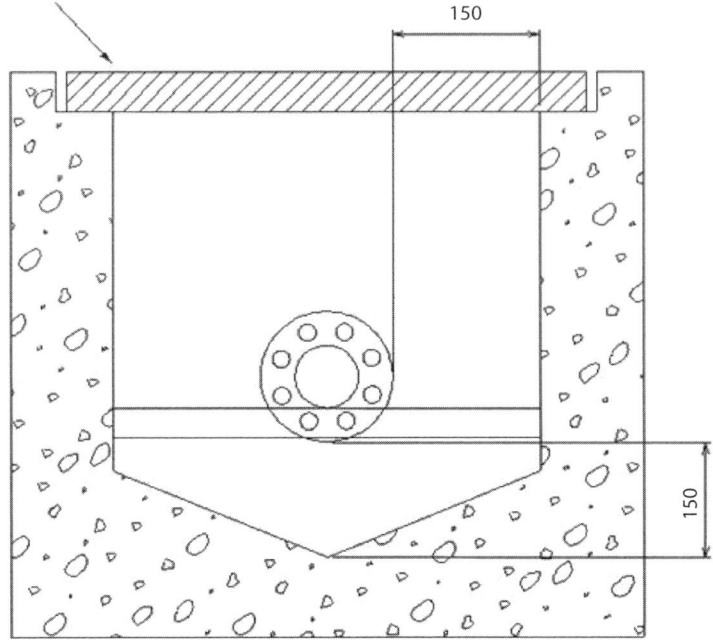

Figure 10–9   *Trenched piping (courtesy of Red Bag/Bentley Systems, Inc.).*

Open trenches require drains to stop accumulation of surface water. The trench bottom should be sloped toward drain points. In this case, the pipe supporting is by means of angle steel or I-beams set into the walls, allowing the bottom for free drainage to nearest drain point.

This method allows drainage of a trench by a minimum number of drain points between each pipe support, as would be the case of solid concrete pipe supports built up from the trench. Before proceeding on trench drainage, check with the coordination procedure and the civil engineers for the water table level.

10.5 Safety Precautions

Most trenches have either a cover of concrete slabs or a grating. Where flammable liquids are carried in trenched lines, a fire break is provided at suitable intervals along the trench and at each intersection. This generally consists of two concrete walls 1–1.25 m apart, with the space in between filled with sand. Where highly flammable gasses are carried, the whole trench, after installation of piping, is back-filled with sand. The piping engineer will advise the civil engineers of the requirements.

10.6 Underground Piping

Keep buried piping to a minimum. Generally, only sewer drain lines and fire mains are located below ground. In some cases, due to client or climate requirements, cooling water lines also are buried below the frost line.

With future maintenance in mind, buried lines should be located well clear of foundations and, if running side by side, well spaced out. A minimum of 300 mm clearance is necessary between foundations and lines and between the lines themselves.

Aboveground safe drain tails enter the below ground drain line via a tundish (concentric reducer normally) or, if a sealed system and cooling water lines, by a flanged stub raised aboveground.

Flanged connections should be a minimum of 300 mm above the prevalent grade level. It is advantageous to set a common level for all these takeoffs at the outset of the job. When locating tie-in connections to underground systems, especially from elevated drain points and adjacent to equipment plinths, ensure adequate clearance.

All buried steel pipes should have a corrosion-resistant coating and wrapping.

Deep valve boxes for buried lines should be designed with ample room inside the box for a maintenance person to bend over and use wrenches to tighten flanges or repack valves. Consideration should be given to the use of concrete pipe in lieu of square boxes.

The criterion for a good underground piping design should be ease of maintenance. Piping should be so spaced as to allow easy digging out and replacement of faulty sections; for this reason, never run underground piping under or through foundations.

Index